LANDSCAPE AND GARDEN DESIGN

LANDSCAPE AND GARDEN DESIGN
LESSONS FROM HISTORY

Gordon Haynes

Whittles Publishing

For Brooke and Natasha

Published by
Whittles Publishing Ltd.,
Dunbeath,
Caithness, KW6 6EG,
Scotland, UK

www.whittlespublishing.com

ISBN 978-184995-082-4

Printed by Short Run Press Ltd, Devon.

CONTENTS

PREFACE

This is not just another landscape history book. It is intended as a primer to highlight some of the design achievements in parks, gardens and estate landscapes over the past five hundred years and to show how the techniques used are still relevant to the practising landscape and garden designers of today. The idea for the book derives from a course of landscape design teaching that I was partly responsible for during a period of about fifteen years as a lecturer at Edinburgh College of Art when that institution was academically partnered with Heriot-Watt University. During this time, the syllabus for first-year landscape architecture students included a survey of historical periods. It was not simply a chronological run-through, but attempted to show the interrelationship between the different curriculum studies, specifically how the principles of good landscape design can be traced from examples drawn from history. An essential ingredient in the programme was a study tour undertaken at the start of the third term of the first year, usually around the second week in May when spring had given essential body and depth to some of the places visited. The plan required staff to organise and lead a rational tour of British (largely English) sites of significant landscape importance, most of which would have been introduced to the students during the first two terms. The philosophy behind the itinerary was to visit sites in the order of their creation, as far as this was possible. It seemed sensible to present information in this way, to reinforce the development thread and to avoid unnecessary confusion. The development of ideas and techniques could then be explained in a way which would make sense of history and illustrate the derivation of the basic design strategies applicable to current practice. For me, no other course of teaching seemed quite as relevant or fulfilling as the integration of design and history in this way and it was particularly satisfying to have been part of this programme, given that I rather struggled to comprehend the complexities of matters historical during my own professional education. As a lecturer, I was a student almost as much as my charges. Staying at least one step ahead of them was sometimes a strain in the early years and I'm quite sure this was plainly evident to the student body.

What we all learned from undertaking the study tours was that it is not necessary to spend hours and hours, still less whole days, at the individual parks and gardens visited. The essential qualities of each site, be they the layout, vistas, composition or detail, can be appreciated in a very short time, usually no more than two hours, before moving on to the next site on the itinerary. This is certainly not to say that I think two hours is the most anybody needs to stay at a historic garden – it merely evolved as the optimum period for realistic study as part of a comprehensive eight to ten day tour. Of course, to be faced with limited time helps to concentrate the mind. Two gardens a day was the usual plan, often interspersed with other relevant studies such as visits to sites of ecological interest and to cities such as Oxford, Bath or Lincoln, where dramatic townscapes were used to demonstrate the spatial studies element of college work, supplementing the garden sites.

At college and later as an educator I had been very keen to find a single reference work which might have detailed both historical sequence and design significance, but I was unsuccessful. Even in the intervening decades such a volume hasn't emerged. There are of course numerous books devoted to garden design history but they tend to single out particular periods or deal with overviews without explaining the design techniques. Here, I have attempted to present concise information in useful, clearly identified sections and subsections that can be accessed at random. I hope it works.

This, then, is not a record of any particular trip, but rather a prescription for an ideal one. I am indulging in the luxury of more time than any of the university tours had available, attempting to make this a comprehensive study. Staying at youth hostels was a necessary feature of those tours. We used them to allow the limited funds to stretch further, but the ideal tour would involve staying at the houses of the gardens, so I invite you to let fantasy take over in your wanderings through this particular Grand Tour of Britain's landscape and garden heritage.

INTRODUCTION

The history of garden and landscape design over a period of five hundred years is necessarily a complex one and is also difficult for the student of this subject to put into any kind of perspective. It may help at first to break down the period into very simplistic categories and to imagine a similarly basic classification, within which the detail can then be investigated. The reality is of course very complicated and doesn't fit neatly into chapters of a hundred years, but a start must be made somewhere. The very broad overview might look something like this:

Sixteenth century	Cellular, complex, lacking a strong masterplan
Seventeenth century	Geometric, controlled and heroic
Eighteenth century	Idealised, naturalistic landscape
Nineteenth century	Ostentatious display and the return of geometry
Twentieth century	Cellular, eclectic and a celebration of plants

Dates cannot be avoided in any historical study and this is no exception. The extreme generalisation that I've suggested above may just be of more use than the specific dates that are quoted within the body of the book. Whilst every effort has been made to be accurate with the dates of the designs, it should be remembered that gardens and landscapes often had extended gestation periods during which fundamental changes regularly took place. Textbooks offer different dates for the creation of particular gardens, some of which might actually be correct. In the context of this book it seems much less important to focus on date accuracy than on the legacy of what was achieved, but I have nevertheless committed to the dates that I believe in.

I feel in good company with the idea of taking entire centuries and giving them labels. In 1938 Christopher Tunnard's Foreword to his *Gardens in the Modern Landscape* had this passage: 'The great ages of garden art were in Italy the sixteenth, in France the seventeenth, and in England the eighteenth centuries respectively. The nineteenth century debased all these traditions to a common level and created a medley of styles'. His simple view approximately mirrors the detailed structure of this book, but only up until the twentieth century, which must appear excessively complicated to all those who scan the contents list. In defence, I can only say that each section heading represents what I believe to be a distinctly different way of making gardens. In another hundred years and with the benefit of a distant view, the picture might seem to be less complex.

I have tried to avoid the use of the terms formal and informal in describing gardens and landscapes, where geometric and irregular would be more accurate. For the same reason, 'architectural' is not a word that sensibly applies to plants, so has also been avoided.

For the most part, the history of garden and landscape design has been closely related to that of country houses, an idea that started in Britain about five hundred years ago with the Tudors. The wealthiest house owners in Victorian Britain enjoyed their gardens, but it was only later in the twentieth century that gardening and garden design became a serious concern for the population as a whole.

The design of gardens and landscapes can be and is taught at college and university level, but students don't just wake up one day and think 'I want to be a landscape architect', without there being some kind of trigger – a curiosity in the disposition of the various elements and features around them which serves as encouragement to investigate the technical aspects of space making and beyond that into the realm of creativity.

For me, an interest in environmental design emerged at the reasonably tender age of 14, when my school organised a study visit for geography students to central Switzerland. It was more like a summer holiday really, involving cable cars, funiculars, paddle steamers, picnics in the high Alps, glacier visits, and so on. I discovered that the Swiss Travel Service (STS) produced wonderful bird's-eye view annotated panoramas that covered the whole country and I acquired them, studied them in detail and have them yet. The exaggerated relief of mountains, gorges, lakes, river floodplains, lacustrine deltas, glaciers and the like makes central Switzerland a natural study location for anyone with an interest in physical geography but it was the morphology, the interlocking volumes of space, the sense of expectation or surprise as a ninety-degree turn in a lake was negotiated (out of cool shade into hot sun), the urgency to see it all and moreover to experience it all that I remember most from that visit. Our base was a tiny hamlet halfway up the Rigi, the same mountain immortalised by Turner in several paintings made on his tour of the Alps. From the summit, the whole natural composition could be comprehended and a plan for discovery worked out. High ground is always good for this. Across the lake from here is a near vertical, north-facing cliff that projects like a finger almost dividing the Vierwaldstättersee into two. This cliff, the Bürgenstock, had to be conquered. Starting from the lake's north shore, a paddle steamer (since replaced by less exciting diesel vessels) makes the short crossing to a landing point barely large enough for a ticket office and a jetty. As this is approached the steamer is plunged into deep shadow cast by the looming rockface. A funicular railway makes light work of the climb, taking seven minutes and delivering the intrepid traveller to a rockface path that is not for the faint hearted. At the end of the path is the Hammetschwand outdoor lift, built in 1905, some 152 metres of freestanding terror that zips you out of the shadows and up to a Meccano bridge that must be crossed, carefully, to reach dry land again. The whole sequence is exhilarating and offers extreme landscape experiences which are completed by a short walk through woodland to a cliff-top hotel. Walk through to the back and see, framed by a picture window, the entire route from the Rigi, and more, revealed in a single plan-like view. A measure of the graphic skills of STS artists is that the reality unfolds exactly as predicted by their drawings. It is more than great views, with water always in the middle distance. It is a complex interconnecting of spaces with ridges plunging into the lake and emerging on the other side like a gateway to the next bay. There is also expectation, concealment and revelation, beauty, sublimity and a deep realisation that nature is good and that a little manipulation of exposure to it can enhance the experience.

A few years later, in the studio at college, I took every opportunity to include high ground, low ground, water, bridges and lookout points in the imaginary schemes that sprang from my drawing board (there were no computers in ancient history). Later still, in practice, the same urge to create landscapes with drama applied. It wasn't that I'd been to Versailles and Blenheim and thought 'How can I do that?' Nor had I properly appreciated that landscape architects were far more likely to be maximising the number of car spaces that can be accommodated at a supermarket than being involved in any kind of glamour project. Nevertheless, the order and sensibility that is required in the laying out of a car park draws upon the same thought processes that designers of the great estates and parks of the past had to grapple with:

- Is the solution the best that can be achieved?

- Does the solution relate to human scale?

- Am I leaving the landscape better than when I found it?

The techniques necessary to achieve these goals are the same too.

This book is about the pleasure garden, not the kitchen garden. It's about how we design outdoor spaces to make us happy and how the appropriate formula changes from one generation to another. Generally these spaces are private and are attached to houses, but we also find recreation spaces in the public domain: in parks, country parks, scenic drives, long-distance footpaths, and so on. Design plays a part in these areas too, though it tends to be either very diverse or pattern book repetitive. I have a new house on a small courtyard development surrounded by a hillside farm. The opportunity to be creative with the modest-sized garden is somewhat limited and the fierce winds that blow from all quarters add a factor of difficulty, demanding that any gardener first establishes a degree of shelter, and in so doing reduces the ground available for the garden. Luckily, I have a readily available and truly spectacular landscape to enjoy as soon as I step outside my front door, as I do daily for my morning constitutional. In the big sky, hooded crows mob a red kite, this time chasing it off as they usually do. Passing a row of Victorian farm cottages and then the splendid Georgian farmhouse, my walk takes me onto a long, straight, single track road that leads eastwards. Immediately a wide view opens up to my left and widens further over the next half mile, to include the entire outer firth. Halfway down the hill are the chimney stacks of the big house, surrounded by dense tree planting. In the foreground are cattle and rare breed sheep, happily grazing the steep meadow. The small four-car ferry might be chugging across the narrow passageway between the cliffs and a cruise liner is likely to be steaming into port having spent the night making its way north from Queensferry or south from the Faeroes to deposit up to three thousand people 'doing' the Highlands in a day. Semi-submersible drilling rigs are parked in the firth undergoing maintenance before being towed out again to a North Sea prospecting zone and shuttle tankers are regularly negotiating the deep water channel that is less than a kilometre wide to berth at the oil terminal. It is a very dynamic scene. To the right a flock of greylag geese darkens the skies having been spooked by some unseen disturbance. The road becomes a track and heads into a wood of ancient sweet chestnuts, beeches and Scots pines that crowd over the headland. The view has gone and been replaced by low shafts of sunlight puncturing the deep shade. The going has got a little more testing too as the track climbs to the car park at its end. From here, above the wood, the views open out again and the panorama now includes the next firth to the north and, on a clear day, a Charles Barry castle. A great spotted woodpecker drums away and a pair of goldfinches flit amongst the thickets. A little way further on and the bastion walls of Fort George, marking the northern limit of the Great Glen, come into distant view. Back at the car park a cliff-side walk called the Hundred Steps leads down to the town and provides excellent opportunities for spotting local pods of bottle-nosed dolphins that frequent these waters and delight in riding the bow waves of incoming shipping. If I'm lucky, a startled roe deer will leap across my trail and, depending on the season, I'll be able to fill a bag with chanterelles for dinner. This rich experience is made possible by there being a road and a path, which incidentally has rather more than a hundred steps, in the right place. As it is entirely in the public domain, the drama is also free. Landowners of the eighteenth century spent their fortunes trying to improve their estates, employing designers to create their own versions of a perfect landscape but sometimes falling well short in the areas of interest and incident. It is a difficult task and not everything can be planned for. Having this particular walk immediately accessible was a potent factor in choosing where we now live.

I have described two landscape experiences, both of which have been enabled by various forms of transport or simply by the provision of access. Access is the key. Roads, footpaths, carriage drives, terraces – all allow access in a controlled way within ninety-nine per cent of the gardens and estates discussed here. Control is in the designer's remit, but there should always be an allowance for serendipity, for those things that are unexpected delights and which tend to stick long in the memory. The process of landscape design is a complex one and cannot be conveyed simply by lines on paper. Lancelot Brown's plans look dull because they don't show the folds in the topography, the light bouncing off his proposed lake in the middle distance or the sheep, deer and crows in the field beyond the ha-ha. Repton did better but was still only able to show static pictures, which may or may not have been achievable on the ground. To walk along the maze of allées at Bramham Park, high beech hedging on both sides and arching branches overhead, might seem a little sterile and intimidating, as it appears to be on plan. However, on a quiet spring morning when the air is still, daffodils are blooming en-masse and the sun is on your back, there is something almost other-worldly about it. The experience of landscape is as much about mood, smell, sound and feel as it is about space, proportion and scale and a designer is a genius indeed if he can plan for and then create all these aspects to the satisfaction of his client. What he can do is to try to set up the right conditions for the serendipitous to happen.

Much of landscape design and of larger garden design is involved with creating a linkage of spaces that produces sequential experiences. It starts with defining boundaries: a wall, a fence, a hedge. Decisions about total or partial enclosure will be made and will affect the height of the boundaries. Will there be open areas to balance shaded ones? What trees and plants can be supported by the soil and microclimate? Will specific activities be catered for? These questions and many more will have to be answered before a credible plan emerges. In some way, all the questions and answers – the interrogation process of design – will relate to human scale. Where is it possible to walk and sit, what can and cannot be seen, what scents and sounds contribute to the experience, how will the maintenance of lawns and borders be managed? With a house, a door opens and a door closes as one room is left and another is entered. With a garden it doesn't have to be as straightforward as that. It rarely is. The American landscape architect Garrett Eckbo defined the nature of his work thus: 'Design is a problem solving activity and at times an art producing thing'. Somewhere in the minds of all the creators of the parks and gardens described here was the desire to produce places of both utility and beauty. Some succeeded more than others and a good helping of vanity has often helped to drive the process.

Some design principles should be outlined first, starting with proportion. Euclid (325–265 bc) was the first to record a definition of the 'golden ratio' of proportion, a device that architects, artists and designers have used ever since. In 1509 the Italian monk Luca Paciola studied proportion in depth and defined the golden ratio as a divine proportion. Le Corbusier used it for his Modulor, a system he devised for architectural proportion. Essentially, the golden ratio, or golden section, concerns itself with only two lengths, a and b, where a+b is to a as a is to b. Another way of expressing this is in the way it corresponds to a rectangle, where one side is to the other as 0.618 is to 1, or as 1 is to 1.618. Proportion is as important in landscape design as it is in balancing the work of artists and architects. As the golden section is universally applicable, garden designs that employ it will achieve a certain harmony that is absent otherwise. Proportion in plan and elevation is only relevant in relation to where the space is being viewed from, or where the designer allows access. Because of the way the human eye is constructed, the facade of a house can only be properly appreciated at a distance of about three times its height, or at an angle of 18°. Further away, it will gradually lose its individuality, but from closer, for instance twice the height (27° viewing angle), the whole can still be seen but it no longer appears in its setting. The appearance of the house in its setting was a vital element for the eighteenth-century landscape designers to work with, although it is probable that their designs were more instinctive than calculated. The fifteenth-century Italian polymath Leon Battista Alberti determined that the ideal breadth of a square should be three to six times the height of the surrounding buildings so that from its centre there would be at most a proportion of 1:3, or a viewing angle of 18°. His formula can apply equally to garden spaces which, if they are designed with an exaggerated proportion, can seem to be either vacuous or oppressive. The norm that many landscape designers accept is 1:4, a ratio that achieves a harmonious balance.

Aesthetics mattered more than utility to many of the architects of the seventeenth and eighteenth centuries. The French architect Augustin-Charles d'Aviler considered: 'The beauty of a building comes before domestic economy and therefore the organisation of space must follow the rules of fine decoration rather than the practical needs of the family.' As an example, François Mansart, working at Maisons, harmoniously distributed the fenestration without a rational connection to the interior spaces. Fortunately, this design policy cannot translate easily to garden works, the nearest parallel perhaps being the construction of sham buildings for picturesque effect.

Scale is an abstract term measured in simple terms, such as large, small, intimate. It depends upon enclosure elements, so at one extreme if there is no enclosure or boundary there is no scale because there is nothing to relate to. If a comfortable space is required, the proportions outlined by Alberti might be followed, but the manipulation of boundary heights (buildings, walls, fences, hedges) brings different characteristics to the design. Spaces can be described as open or enclosed depending upon the proportion of space width to boundary height. Enclosure can be perceived even though there might be only slight physical barriers to a space.

With garden and landscape design the natural potential of the site will always be a factor. Briefly, this pertains to the climatic climax vegetation and the proposed planting regime in the design. The further away from the climatic climax, the greater the effect on the continuing maintenance and management regimes. Planting structure is arguably more important than detail, for without the structure the detail may not survive. Some garden designers never concern themselves with the fine detail of planting plans, preferring to set up the conditions for the garden owners to express themselves within a well-proportioned framework. In the natural order, there are four definable layers or zones of planting which the designer can choose to emulate or select from. These are the field layer, the herb layer, the shrub layer and the tree canopy. Selecting anything other than a representation of the natural order will result in variations of artificiality. Trees are particularly useful as a scale-measuring device and deciduous trees are important as they help to mark the seasons. Flowers, though important in many gardens, are temporary and must be planned as a seasonal decoration rather than a structural component. Planting is very versatile. It can act as a screening device and a directional aid, as well as a sheltering agent and to provide enclosure. A well-tended lawn often gives the impression of the space it occupies being larger than it is. A clean-edged boundary between a lawn and a border can compensate for almost any unplanned incursion such as weeds.

Manipulating the land, rearranging the contours so that they assume artificial or naturalistic form, has always been a basic option for garden designers. Although labour intensive and therefore expensive in past centuries, it is now relatively cheap and quick to achieve. Landform has a tendency to survive longer than any planting, building or structure, so should always be handled with great care. Like tree planting, landform can effectively describe mass and void. It affects perception of space too: a downhill view will generally appear to be longer than an uphill one.

For many people a garden lacks life without water and designers have recognised this since the earliest times. Water plays on the senses. It reflects light and magnifies the visual range of a landscape. It modifies temperature and, when in motion, contributes to the audible range. It can dazzle and offer repose. The treatment of water, particularly the water's edge, can make the difference between success and failure of a design – nothing betrays inaccuracy of detailing more than a sheet of water, which always finds its own level. Water is not a maintenance-free surface. If it leaks, the design stands to fail and its escape could lead to collateral damage of neighbouring property.

A design will need unifying elements to make sense of its composition. As mentioned above, these can include paths or roads which offer an invitation to follow and therefore determine the sequence of spaces to be experienced. The Arts and Crafts gardens of the Edwardian period often used the same materials for walls and paving in the garden as were used in the houses, thus bonding the one to the other. Other devices that invite exploration include the sight of light through dark and framed views. Designers must always be aware that the eye exaggerates vertical features, which is why photographs often don't trigger the same response as that which prompted the taking of them.

Colour can be either inherent to the materials being used or applied to materials and can be perceived either by natural daylight or artificial light. In each of the combinations it will be different, a circumstance complicated by the fact that daylight itself is constantly changing, modified by atmosphere. Aerial perspective, which is the progressive diffusion of light rays with distance, dulls and uniforms colours. A blue/grey haze is the usual perception of distant objects in Britain, due partly to the humidity in the atmosphere. As weather changes, so the light conditions change and colours are thrown into highlight and shadow as well as all grades between. It is nearly always the case that the extremes of highlight and shade provide the most dramatic prospects. Intensity of contrast can diminish colour to the point where it is virtually lost and shadows, with or without 'light through dark' situations, on one side of a space will make it seem larger. The orientation of the light source can brighten or dull colours, particularly on materials with textured surfaces. Consequently, walls and hedges in south-facing gardens should be carefully ordered so as not to adversely affect other elements of the design. Colour and pattern can be used effectively to signify ownership or change of use – the greater the contrast the quicker the response. Contrast, whilst dulling colour, can be particularly effective in presenting form and texture in silhouette, for instance trees against the sky where the winter branches can be distinctive and influence the character of a space. A garden designer should always have a colour wheel to hand when planning contrasts and harmonies.

These are some of the parameters that garden designers work within, their best designs being those which provide the solutions for the required uses but also look good and feel good to be in. Eckbo was happy to relegate art-producing as a potential spin-off in the process of garden and landscape design but art is becoming increasingly undefinable. Another designer, Lord Holford, had a slightly different view. Referring to landscape design, he suggested that it 'is a process of selection among a host of variables to produce unity, impact, symbolic significance and permanent cultural values'. Gardens and landscapes are, as we shall see, part of a living and changing process in which the design must address the fourth dimension and anything that can successfully achieve that is at least very clever.

We start where we should, at the beginning.

1

EARLY RENAISSANCE EXPERIMENTS
1500–1688

Principal designers	Extant examples	Principal components
John Needham	*Hampton Court Palace, Surrey*	*Topiary*
Salomon de Caus	*Kenilworth Castle, Warwickshire*	*Fountains*
Isaac de Caus	*New College, Oxford*	*Parterres and knot gardens*
John Rose	*Edzell Castle, Angus*	*Terraced walks*
John Tradescant, the elder	*Hatfield House, Hertfordshire*	*Grassed and low mazes*
André Mollet	*Ham House, London*	*Mounts*
John Slezer and Jan Wyck	*Oxford Botanic Garden*	*Enclosing elements*
Sir William Bruce	*Powis Castle, Powys*	
	Kinross House, Perth & Kinross	

Garden histories usually identify particular styles with the monarch of the day, sensibly so because it was they who tended to set the standards, or encouraged them at least. This review starts in the sixteenth century, a period which has left us with very little of a physical heritage to study. It was the Tudor century and can, very approximately, be divided in half, with Henry VIII occupying the first half and Elizabeth I the second half. Strangely Elizabeth usually gets her own subset but Henry doesn't. Edward VI and Mary I occupied eleven years in the middle but we can easily disregard that. The Jacobean and Caroline eras occupied the first half of the seventeenth century, again dividing it almost equally. The Commonwealth of 1649–1660 ended with the Restoration of Charles II, who died in 1685. Broadly speaking, Henry VIII was a builder of palaces and gardens, Elizabeth I was not, but the noblemen of her reign built gardens to express their wealth and generally to impress. Charles II was the only seventeenth-century monarch to take the arts, architecture and science seriously until William and Mary acceded following the Glorious Revolution of 1688. James II (1685–1688) had no time to think about gardens and, under Cromwell, there was more destruction than construction. The House of Orange introduced a Dutch flavour to the Grand Manner and is discussed in the next chapter.

The Renaissance took a while to spread north from Italy and when it did so it manifested itself in garden design only for the extremely privileged and wealthy. Before the Grand Manner took hold in the late seventeenth century, a limited number of significant gardens were made in Britain which developed from the continental influence, moderated by a home-grown medieval tradition. They were normally attached to palaces or castles but on a smaller scale they also started to appear as courtyard extensions in university colleges. Most domestic-scale gardens were probably practical rather than artistic in any way.

There were subtle differences between the gardens of each era. Tudor gardens were never intended to provide a setting for their houses or to enhance the building works in any way. The house and garden were not bound together as a unified whole but were arranged in a way that best satisfied the growing of flowers and the display of ornaments, usually together and often in equal proportions. They were subdivided into visually enclosed spaces, contrast between one and the next being more important than any idea of a steady transition. Just as in the nineteenth century, when writers and architects alike set out rules and principles for garden design, there was no shortage of advice available for garden makers. Richard Surflet translated a French book, published as *The Countrey Farme* in 1600, in which he referred to the 'inward proportions' of gardens, listing several features and how they should be divided, without any reference to a relationship with the house. Francis Bacon (1561–1626) must be mentioned at this point. A major figure in a number of fields including philosophy, the law, writing and science, he had views on everything and expounded them whenever he could. His *Naturall Historie* was published in 1627 when he could no longer be called upon to justify anything, perhaps luckily for him. It was full of superstition presented as fact, which, given that he is recognised as the father of the scientific method, raises some fundamental questions. He offered advice on horticulture which must have had gardeners pulling their hair out when they discovered that reality bore no relation to learned understanding. What were they to believe when Bacon declared that most trees have no flowers and that elm and box have neither flowers nor fruits?

Tudor gardens were about nothing if not enjoyment. They were packed with ornament, colour, illusion and trickery. They were places to play. Nan Fairbrother, in *Men and Gardens*, summarises Tudor gardens as having 'borrowed ideas from Italy, but … they speak Italian with an unmistakable and unrepentant English accent'. Allegory played a large part in their detail design, particularly the earlier ones. Religion was at the heart of this. Gardens were recognised as a mirror of heavenly order where the natural cycle was witnessed every year and with wild nature as the contrast beyond the walls. The representation of the Apostles in evergreen topiary is a major part of this allegorical theme.

Jacobean gardens appear to have been elaborate, complex, fussy. In 1624 Sir Henry Wotton, a respected diplomat, assessed the Jacobean gardens of the early seventeenth century as being 'a delightful confusion'. Francis Bacon described ideal gardens as having all-year-round interest but specifically disliked knots, topiary, statues and stagnant pools. Caroline gardens reined in the confusion a little and tended towards the Italian Renaissance in character. There was axial symmetry of the house and its related garden and a terrace usually closed the composition at the far end from the house, from which elevated position the house and garden could be viewed. The opportunity for viewing often extended along both sides of the rectangular garden, effectively enclosing it and placing it at the lowest part of the site. Descriptions survive of these gardens but only one site has survived, at Bolsover Castle in Derbyshire, recently restored but sadly short of interest, being a simple walled enclosure of grass with a fountain in the centre.

In the mid-seventeenth century, the immature French style of garden design, as practised by André Mollet and Boyceau de la Barauderie, already had a definite fixation on axial symmetry. Boyceau, in a book published posthumously in 1638, declared 'All things, however beautiful they may be chosen, will be defective if they are not ordered and placed in proper symmetry'. The Restoration of Charles II fuelled the fashion for French-style gardens hand in hand with a building boom of new Baroque country houses. The Tudor, Jacobean and Caroline styles, many gardens of which Cromwell had done his best to destroy, hung on precariously whilst Mollet was thinking big. The old-style enclosed gardens began giving way to those that looked outwards, in some cases even introducing the wider landscape as part of the composition.

Design detail

Tudor gardens included almost any kind of feature as long as it contributed to their decoration and therefore enjoyment by the owners and their guests.

Knot gardens and parterres were always located well below eye level so that their intricate detail could be properly appreciated. Upper windows of the palaces and castles were related to the gardens which were also furnished with elevated terraces for those perambulating in the gardens. Knots illustrated the artistry of the gardener and were to be admired. It was a 'Look, don't touch' approach to garden design. Patterns were often very intricate, sometimes allegorical, and involved apparent interweaving of clipped hedging, typically box, but also herbs such as rosemary. Open knots were sanded between the hedging, closed knots were filled with single coloured flowers.

Knot and parterre gardens are necessarily made on level ground so the mount was developed partly so that a better view of the intricate patterns could be gained. Because many gardens were enclosed by walls for protection, the mount also allowed views out, from a safe position. They were common features in Italian gardens from the mid-fifteenth century, which may be the origin of British examples. Many larger mounts also had garden buildings of some kind on their summits, which were reached by a spiral path edged with hawthorn. Mounts and terraces did the job outside that upper windows of the house did inside. Some mounts were constructed of timber and were understandably transient features. Earthworks have a habit of surviving where other elements fade and so it is with earthen mounts.

Punctuation points in the cellular garden arrangements were made with arbours, obelisks and fountains. Aviaries were also popular, as was 'carpenters' work', which included trellises and frameworks, often archways, for training climbing plants over. André Mollet introduced scythed turf as a component of British gardens.

Topiary was an ever-present element. Anything was game, be it geometrical forms, men, animals, chess pieces, ships, flowers, and so on. Clearly a skill base was developed in these years, comprising many types of craftsmen as well as gardeners. To maintain them in peak condition would have involved almost constant attention, with the gardeners being temporarily ushered away when the gardens were being used for entertainment.

The Restoration period became known for the popularity of lime avenues, although Hadfield (*A History of British Gardening*) says that they appeared much earlier in the century.

Hampton Court Palace

Needham, sixteenth century; Rose, London, both seventeenth century
From 1532

Henry VIII acquired this estate in 1525 and in 1532 started the process of constructing gardens on the south side of the palace. John Needham was the Surveyor of the Office of Works during much of Henry VIII's reign and was therefore responsible for the gardens which Henry commissioned.

The Privy Garden had parterres and high-level terraces to view from. There was also a Mount Garden close to the River Thames, made in 1533–1534, and three ponds in a Pond Garden closer to the palace. The king was particularly fond of heraldry and sundials. Heraldic beasts, carved in wood, and twenty-two sundials occupied the Privy Garden alone. The maze is reportedly the first hedge planted maze in the country and was added to the gardens on the north side of the palace in 1691 (but see below at Nonsuch). The original hornbeams were replaced several times, most recently with yews in the 1960s. It was in the middle of that decade when I first saw the maze and I remember being distinctly underwhelmed, both by its proportions and by its general rattiness. There is so much more awareness of our garden heritage now that such a public display of decay is unlikely to be allowed in the future.

Recent restoration of the park has involved replanting of the great lime avenues of the Long Water, the central axis designed by John Rose and later incorporated into the patte d'oie designed by George London and defining the zone of the deer park. Part of the deer park, or Home Park, is now a golf course. The Privy Garden was restored to a plan of 1710, as confirmed by a measured drawing of Inigo Triggs almost two hundred years later, but the Mount has never been reinstated.

The park and gardens at Hampton Court were significant in their time for their scale as well as their design. There was nothing else in Britain, until Nonsuch, to match them. For many years now the semicircular parterre garden on the east front, also known as the Great Fountain Garden, has been criticised for its visual disconnection with the Long Water, the avenues and the deer park. The English yew trees, which are pruned into broad cones from just above head height, block the views across this space and beyond. A hundred years ago they were no less of a barrier to views but were free growing. They replaced London's original low box hedging in the early 1700s, Queen Anne preferring wide lawns and topiary to the many fountains and parterre work. It is also rumoured that she didn't like the smell of box. What is claimed to be the longest herbaceous border in the country separates the Privy Garden from the Great Fountain Garden.

The consequence of all this is that Hampton Court today faithfully represents only one feature from its earliest garden design, albeit restored, namely the Privy Garden. It is unlikely that any changes will occur to the conical yews in the near future so the awkward visual relationship mentioned above will continue to provide critics with ammunition.

NONSUCH
After 1538

Not far away from Hampton Court, this was a very important royal palace and garden. It had hedges, fruit tree avenues, a knot garden, a bowling green, a tennis court, a maze, a mount with a banqueting house on top, trellis work and water trickery in profusion – an idea borrowed from Italian Renaissance gardens. In short, the garden had everything. The maze might originally have been a low one, as was the fashion, but in 1599 Thomas Platter the younger, a traveller from Switzerland, wrote of the maze that it was 'thick and high enough to prevent cheating'.

The palace, which Henry VIII built from scratch to be even grander than Hampton Court, and which was Elizabeth's favourite palace, was demolished in 1682. Nothing remains either of the gardens which, when they passed into private hands, were anyway considerably altered.

KENILWORTH CASTLE
1575

In May 2009, English Heritage opened the garden here after a £2 million restoration that was based upon a detailed description made at the time of its construction. Much professional and public criticism has followed the opening, mostly based upon the scale of the detailing but also on aspects of the layout. It is generally held to be unattractive, contrary to the spirit of Tudor life, which engaged artistry of all types and was far from dull. Kenilworth was the seat of Robert Dudley, who entertained Queen Elizabeth here several times. On the final occasion, in July 1575, there was a nineteen-day-long party for which these gardens were created, specifically as a Privy Garden for the queen. The garden didn't persuade Elizabeth to marry Dudley and its design doesn't convince everyone today. However, the composition is likely to be broadly correct so it offers a window on the past, even if the glass is slightly fogged. Arbours, obelisks, a central fountain and a large aviary all feature in the layout, which is viewed from a terrace three metres above the garden and between it and the castle. The restoration has used only plants that were available in 1575, using flowering species that are at their peak in July.

No detailed drawings survive of new gardens made in Elizabeth's reign. Theobalds, in Hertfordshire, and Wollaton, in Nottinghamshire, both from about 1580, were larger than Kenilworth and their general layout plans indicate a strong French influence, which is not surprising given that many gardeners in Britain were French too. Subdivisions were squared within a square or rectangular framework.

NEW COLLEGE, OXFORD

From 1594

New College has the oldest quadrangle in the city (1386) and some very fine railings erected in 1711 to designs by a pupil of Tijou, who was responsible for the fine screen at Hampton Court Palace. Chief amongst the garden attractions is the mound, or mount. Now heavily planted, which is not quite how it was designed to be, the earthworks are nevertheless still reasonably intact, despite the original stepped arrangement having been evened up to a regular gradient. It is therefore a rare surviving example of this feature which was devised to allow views out from the walled enclosure of the garden. The garden was not originally laid out for recreation purposes. University records suggest that it might have assumed a geometric plan by about 1530 and that the mount construction was started in 1594 and not completed until 1649 (really, fifty-five years?), when it was 'perfected with stepps of Stone and setts for ye Hedges about ye walke'. For setts, read quickset or hawthorn. Part of the medieval city wall and a fine herbaceous border are the garden's other attractions.

Wadham College had an unplanted mount until the mid-eighteenth century which acted as a base for a statue of Atlas.

EDZELL CASTLE

1604

Early sixteenth-century Edzell Castle is now a ruin but the garden, or pleasance, with its twelve foot high heraldic walls and garden house built by Sir David

Lindsay, Lord Edzell, in 1604 has survived. As with so many of Scotland's castles, Queen Mary once stayed here, in this case holding a Privy Council meeting in August 1562.

The site is in a hollow at the confluence of two rivers. The castle's tower house was built primarily as a fortress and even the walls of the pleasance were built with defence in mind. The whole garden had to be built within the constraints of an old moat, which was swept away by a flood in the middle of the eighteenth century and is a rectangle of only about sixty by fifty metres. The garden layout is an imagined reconstruction, dating from 1932, as no detail of the original layout has survived. A contemporary writer noted: 'The castle or palace of Edzell is an excellent dwelling, a great house, delicate garden, with walls sumptuously built of hewn stone, polished, with pictures and coats of arms in the walls.' In summer, blue and white lobelias are planted in recesses of the wall, the stone of which is deep red, reproducing the heraldic device of the red, blue and white fess-chequy coat of arms of the Lindsays. Suitable plants were not available at the beginning of the seventeenth century so the original design for decorating of the recesses is unknown. Whatever turmoils of earlier centuries were brought upon Edzell, it is now a very peaceful place, consistent with it being ruined of course, but something of the spirit of the early seventeenth century just following the union of the Scottish and English crowns is retained, as is most of the extraordinary craftsmanship that was lavished on the walls, which are still possibly the most sculptural and attractive of any in the country.

HATFIELD HOUSE

De Caus, Tradescant
1607

De Caus was involved with the later stages of laying out the gardens at Hatfield House, with John Tradescant providing the planting design. The house stands on high ground and the garden descends from it through shallow terraces. Many of the features typical of the age were included in the garden, which is currently being restored. There was a very large vineyard in this, the first garden to introduce the French Renaissance design influence to Britain, half a century before Le Nôtre was to hit his stride. A very impressive north/south axis, which is marked by a close avenue on the north approach and a wide avenue, grassed, to the south, passes through the house itself. Hedged rectangular enclosures, which cluster around three sides of the house, enclose knots of various design and a large maze now sits between the house and an eighteenth-century-like lake to the east. In the seventeenth century the water feature surrounded the house and was bordered with trees.

Although much altered and added to, the general character is what is believed to be seventeenth century, pre-Grand Manner.

Oxford Botanic Garden

1621

Originally called the Physic Garden, this is the oldest botanic collection in Britain as well as one of the smallest. The garden itself, which is part of the university, is approximately square and walled and is located opposite Magdalen Bridge where the River Cherwell flows through the centre of the city and marks the garden's eastern boundary. Within, the land is subdivided into four equal parts with paths crossing at the centre. Of course the detail has changed a little over the centuries but the basic layout survives and it is beyond the walls where glasshouses and offices can be found, out of respect for the original plan. The garden is open to the public and readily accommodates passive use. Not being designed for this, it nevertheless shows that a high wall, to close out the city noise and protect the plants within, some paths and of course the plants (even if set out in rows for scientific study) are ingredients enough to make a much appreciated facility, being a counterpoint to all around.

Garden lovers frustrated at the often bizarre opening arrangements of the college gardens can always gain entry to this most venerable of Oxford's open spaces and admire Nicholas Stone's walls and gateways, still sound after nearly four hundred years.

Ham House

Slezer and Wyck
1671

The first garden here, from about 1610 when the house was built, followed the current French character and had a rectangular layout related directly to the house facade. Later in the century, something approaching the mature Le Nôtre programme was adopted with parterres to the sides of the house and a long axis passing through it from the River Thames to the north to the village of Ham in the south, a distance of over a mile. The oldest orangery in the country is one of the attractions here, as is an icehouse. John Slezer and Jan Wyck, who also prepared plans for, or of, Lennoxlove in East Lothian, were responsible for the layout of 1671, upon which the restoration – which started in 1975 – has been based. The cherry garden, to the east of the house, now has a spectacular lavender parterre and to the south there are eight square grass plats, exactly as drawn but posing an obvious question 340 years after their introduction: Why isn't this ground laid out to elaborate ground level patterning? Certainly the ground immediately to the south of the plats provides a complete contrast in that it contains the wilderness (trees, hedges, statuary) but the Grand Manner would have had parterres and fountains here, not squares of grass. Similarly, in the Île de France, the chief axis would be much wider than the rather mean avenue here and would have been 'carved out' from a deep woodland. The Slezer and Wyck plan gives the impression of representing a transition stage.

LANDSCAPE AND GARDEN DESIGN

As it is today the garden is a mid-1970s National Trust restoration which has reinstated both the layout and detail faithfully, using contemporary sketches, plans and inventories as well as aerial photography from 1949.

Powis Castle

1680

Powis is a defensive castle dating from about 1200. When defence became less of an issue, gardens were developed in the land around, although the topography made this rather a challenging prospect. The castle stands on high ground and looks out over a cliff to the south-east which was excavated in the late seventeenth century to create the closest approximation of an Italian Renaissance garden, specifically the hillside ones, anywhere in Britain. Bodnant Hall (q.v.) is less severe and feels more British than Powis. The castle is a huge slab of pink stone and completely dominates the composition with its castellations and square tower, although grotesque yew topiary on the top terrace makes a bold attempt to compete for visual attention. There are four terraces, the lowest of which has an orangery. Lead statuary, some by van Nost, stone balustrading and arcading all adorn them. There was once a water garden in the valley bottom below the castle and no doubt it would be a truly spectacular location for a Villa Garzoni-like water parterre. William Emes, a contemporary of Lancelot Brown and one who followed Brown's design philosophy, made a landscaped park in the valley towards the end of the eighteenth century. Now the ground is open lawn, fringed with trees and shrubs, forming an unsatisfactory visual relationship with the terraces and castle. Ultimately, this is the great disappointment of Powis. It really needs its water gardens as a counterpoint to the castle and terraces and anything else will always be a fudge. A plantsman's collection of randomly arranged trees with coloured foliage, rhododendrons and magnolias is unworthy of this site, which screams out 'Rescue me'. More recent developments to the south-east offer a range of garden experiences, not least some exquisite hedging and simply clipped yews.

Kinross House

Bruce
1679

In Scotland, the original garden design at Kinross House, Perthshire, planted between 1679 and 1685 by Sir William Bruce, is very much in the French tradition. Axial, symmetrical, walled and subdivided with various parterres, the focus of the garden composition is a ruined castle on an island in Loch Leven beyond the garden's celebrated fish gate. Whilst the planting detail has been changed in an early twentieth-century reworking, the view, space, proportion and scale of the composition remain as a fitting conjunction to the imposing monolithic house. Bruce had done the same thing in 1665 with the view at Balcaskie, in Fife, where the Bass Rock occupies the distant focus of the terraced garden. John Anthony, in *The Renaissance Garden in Britain*, suggests that this feeling for the landscape was far in advance of general practice anywhere else in Britain and prefigured the Landscape Gardening Movement of the eighteenth century. The foregrounds, or the garden element, of designs with axial symmetry are, in Bruce's compositions, less satisfying than the distant prospects and the notion that the estates have been orientated specifically for features beyond their boundaries. One interesting aspect of the work at Kinross is that the garden – its walls, pavings and planting – was started seven years before the house. Bruce was clearly committed to the site and knew that a mature garden was going to take rather longer to achieve than the building of his new house.

Balcaskie was inspired by Vaux le Vicomte but has Italian-like terraces, the detailed planting of which has also been changed over time. The setting and distant prospect remain wonderfully intact.

Other sites

Aberdour Castle, in Fife, has a series of three terraced gardens, laid out in an L shape on plan on very steep grading land below the south wall of the castle. They date from after 1553 and in 1690 an orchard was planted on the ground below the lowest terrace. This has now been replanted and the walls of the terraces rebuilt. Sadly, no details exist of what the terraces originally contained, so they have been turfed instead. Buxbaum, in his *Scottish Garden Buildings*, reproduces a conjectural bird's-eye view of Aberdour in 1650 which offers a flavour of the rich detailing of the period. A fine sixteenth-century beehive doocot stands at the south end of the terraces and a seventeenth-century walled garden of over an acre has very high walls but no vestige of any original content. The terraces were not exactly an original idea because there were others, notably at Barncluith near Glasgow, but few have survived.

At Haddon Hall, Derbyshire, there are terraces with great stone buttressing, similar in character to Aberdour, overlooking the River Wye. In the 1920s a topiary garden with figures representing family heraldry was made as part of the property restoration. The gardens were originally a seventeenth-century creation. This is a massive house with a big garden.

Replanted in the late nineteenth century, the courtyard garden at Montacute, in Somerset, is late Elizabethan (1588–1600) and notable for its fine garden pavilions and elaborately balustraded walls with finials. Visitors have to imagine what it might have looked like originally, especially since Vita Sackville-West (of Sissinghurst) and then Phyllis Reiss (of Tintinhull) added planting designs to the eastern courtyard during the twentieth century. The courtyards are at the rear of the house now but they were designed as an entrance sequence to the magnificent house. It has always struck me as quite a tour de force that such an imposing house can have its entrance orientation altered by 180° without the result looking strange.

A small walled garden in Oxfordshire, at Chastleton Manor House, has a design from 1614, much simplified now but interesting for its concentric circle layout of beds, yew hedges and topiary figures all centred on a sundial.

At Stirling Castle, or rather outside the castle on level ground to the south, is the remains of the King's Knot, dating from about 1625. The king was James VI and I and the fact that gardens were laid out on open, undefended land must be viewed in the context of the peace which had followed the union of the crowns. The King's Knot is now only an elaborate octagonal earthwork within a lower, square earthwork, all centred on a mount. It sits alongside another shallow earthwork of equal size and consisting of a square within a square with cross paths and a central circle. This is assumed to have been a parterre. When the sun is low in the sky, these ghostly earthworks seem to emerge from the field as they cast a shadow. Their novelty value is greater than any conjecture about the design content they once had and their resemblance to some of the works of the late twentieth century (Graeme Moore, Charles Jencks) is evidence that what goes around comes around.

Wilton, in Wiltshire, is now an eighteenth-century park but Isaac, younger brother of Salomon de Caus, made here the first British garden in the complete Italian Renaissance style (more Venetian than Roman), in 1632. Being on a level site, the layout was largely a pattern on the ground, surrounded by a raised terrace. Water tricks abounded. The Palladian bridge at Wilton, added later and for which the garden is well known, was copied for Stowe, Prior Park in Bath and Tsarskoe Selo in Russia.

Pitmedden Castle, Aberdeenshire, has a reconstructed walled garden from the late seventeenth century. The original designs were lost in a fire but it is known that they were to suit with the French style which was reaching its peak at the time. The parterre which is now at Pitmedden has only passing interest for scholars of history and garden design, being a recent planting based on John Rose's designs for Holyrood House in Edinburgh.

The first house at Compton Wynyates, Warwickshire, was a moated manor that was started in *c.*1480 then upgraded in 1515. As well as a moat there were typical Tudor gardens behind the house and a courtyard garden fully enclosed by the building itself. In ruins in the early nineteenth century, some restoration was begun in 1835 and in 1895 the well-known topiary garden was laid out. About half of the moat survives but the topiary has now all gone.

Blickling Hall, in Norfolk, had gardens before the property was bought by the Boleyn family as early as the 1450s, but it was in 1616 that the gardens were brought up to date with parterres, ponds and a mount. Unfortunately this was all swept away in the eighteenth century, partly by Humphry Repton and his son John in 1793. After 1840 the geometry returned and now the gardens have bits and pieces redolent of many eras.

Legacy

Despite there having been a mass destruction of the early designed, Renaissance-influenced gardens, either by overworking or merely through vandalism, the sixteenth- and early seventeenth-century examples became extremely relevant in later centuries as architects, gardeners and landscape architects all sought a return to a sense of order and delight in their gardens. It was more the individual components that were cherry-picked than any sense of the garden as a setting for the house because, as we have seen, that wasn't considered to be important until later. It emerged at the end of the period in the Restoration gardens, which, leaning heavily towards the Grand Manner, introduced axial symmetry of both house and garden. Tudor, Jacobean and Caroline gardens may sometimes have looked a little haphazard in their planning but they could never be accused of being sterile.

Perfectly satisfactory cellular gardens, designed for living in and for fun, gave way to intellectual, mathematically precise designs that aped the Renaissance styles of the continent. In these later gardens the owners of the estates might have been exercising their control over nature but they did so at the expense of enjoying them. Showing off was the order of the day.

It is perhaps ironic that as the slow change from early Tudor to Restoration was seen as progress, it is much more likely that gardens today will contain elements of and be ordered in the Tudor way than anything approaching the Grand Manner. The Edwardian gardens at the beginning of the twentieth century attempted to blend the two styles, with some success, although on a relatively small scale which is naturally unsuited to great gestures. Lutyens used avenues and axial planning as approaches to several houses which had enclosed cellular spaces on the garden front. Great Maytham, in Kent, is one such.

2

THE GRAND MANNER
LATE SEVENTEENTH CENTURY – 1720s

Principal designers/ practitioners	Good extant examples	Principal components
(Le Nôtre)	Bramham Park, Yorkshire	Large scale
André and Gabriel Mollet	Melbourne Hall, Derbyshire	Woodland with intersecting allées
John Rose	(St Paul's, Waldon Bury, Hertfordshire)	
George London and Henry Wise		Statuary and urns
	(Cirencester Park, Gloucestershire)	Geometric plan, usually symmetrical
Charles Bridgeman		
Robert Benson, 1st Lord Bingley	Wrest Park, Bedfordshire	'Canals' and other waterworks
	Boughton Park, Northamptonshire	Parterres
John Erskine, 6th Earl of Mar		Wilderness plantings
	(St James's Park, London)	

The Renaissance spread from Italy into northern Europe during the seventeenth century and whilst travel to Rome and Florence became easier at this time it was the French and Dutch versions of garden design that took a much stronger hold in Britain. André Le Nôtre, as Royal Gardener to Louis XIV, was the main exponent, his breakthrough design being that at Vaux le Vicomte for Fouquet. This notoriously gained him work at Versailles which led to many other palaces, mostly within the Paris basin or Île de France. This was the area where wealth was concentrated and its comparatively featureless landscape favoured a large-scale approach. The proximity of northern France and Holland was clearly a major factor in influencing those landowners laying out their estates in Britain, many of them travelling to Paris to consult with Le Nôtre. He himself was too busy to take on and supervise foreign commissions but nevertheless sold drawings and designs which were then carried back and adapted for British sites. The style was less suited to the Scottish landscape than the English, but this didn't deter William

Adam at Taymouth Castle or the 6th Earl of Mar at Alloa Tower, where designs of great extent were laid out despite the unpromising topography. Adam's avenues and parterres probably sat unsympathetically in the context of the upper Tay Valley and had a relatively short life. In the spring of 1981 I led a student study tour which included Taymouth. Our serious observations were halted at some time during a walk along the sinuous, river-hugging lime avenue to the rear of the castle. We had concluded that the avenue must have been one of the first expressions of the move in taste away from the Grand Manner, although I confess that our reasoning was assisted by a venerable bottle of Talisker 1957. I mention this only to reinforce the notion that even on intensive study tours, history can and should be enjoyable. In their defence, I should say that this group of students went on to graduate very well and to enjoy prosperous careers.

At Alloa Tower the estate layout on the banks of the River Forth was particularly ambitious. Indeed the estate plan of 1710 indicates an even larger estate than Versailles and one consisting of every component which might be expected in a French garden, including a wilderness, thought to be the first introduction of this feature in Scotland. Although Mar spent his life in exile after the 1715 Jacobite Rising and never saw the fruits of his labours at Alloa, John Macky, a spy for the Crown, visited in 1729 and described the landscape in some detail, focusing on its size ('far exceeding Hampton Court or Kensington'), the extensive parterre between the house and the River Forth ('from whence … you have 32 different vistas each ending on some remarkable seat or mountain at some miles distant') and on the wilderness areas which have 'trees for birds and little grottoes'. Sadly nothing remains, although the tower itself was restored in the 1990s and its immediate surroundings have been laid out to reflect some of the original character. Many other gardens were started during the seventeenth century and reworked in later periods, some retaining certain elements of their original design. The garden at Hampton Court Palace was started in the early sixteenth century and has extant features from each of its several remodellings including the Long Water 'canal' by John Rose in 1661. The patte d'oie and avenues designed by George London followed in William III's reign. As discussed later, the Mollets worked on the original layout for St James's Park, London, as did John Rose. The 'canal' was transformed to its current irregular outline by Nash in the early nineteenth century. Sites like this offer particular characteristics for the garden enthusiast to study but are often so complex in layout that they present a confused picture which relates only weakly to a logical composition or masterplan. In this respect Hampton Court remains true to its Tudor origins.

The essence of the French style was an obvious control and domination of nature by man's ordering hand. It follows that no aspect of the design could be allowed to appear random or naturalistic. Layouts were axial to the house or palace with avenues and pattes d'oie emphasising the importance of the house itself as well as focusing on distant features. The design of the layout and its main display should always be clear when viewed from the house or terrace on the garden front. Some slight variation in topography was desirable as it allowed the development of raised terracing to overlook parterres and for water features like 'canals' to extend into cascades. Gardens of French palaces were designed to accommodate the entertainment functions of court and consequently only came alive when populated on such occasions as fêtes. In Britain the entertainment element was also important but the style was adapted to suit all sizes of gardens, whether attached to town properties or country seats. It was after all only royalty, the aristocracy and those with considerable wealth who were in positions to commission these gardens, there to undertake their social and official duties. *The Theory and Practice of Gardening*, 1712, translated by John James from the 1709 French original (d'Argenville), describes in great detail the methods used by Le Nôtre and became the bible for all estate improvers.

Design detail

Gardens of the Grand Manner were designed to impress. The landowner could stand on his terrace with his guests and look down on exquisite parterres perhaps with fountains to animate the foreground. Without moving, the party could lift their gazes to the distant horizon framed by a number of radiating avenues, the central one containing a water body of some size and beyond that a monumental sculpture silhouetted on the skyline. All of this involved civil engineering work on a grand scale and, the randomness of the surrounding landscape being excluded from view, declared with conviction that the owner was in total control, even of nature itself. Not all was immediately visible though. Despite appearances, gardens were far from being one dimensional. The woods that flanked the main vista contained hidden delights, discovered only by turning one's back on the main event. Here would be outdoor theatres, labyrinths (mazes), bowling greens, quincunxes, fountains, a wilderness, and so on. Access to them was by intersecting rides which cut through the woods and sometimes connected to ronds-points. To elaborate a little on two of these features: a quincunx describes geometrically arranged trees, typically in the form of multiple fives, as on dice. In the larger examples they form an offset grid. The wilderness was an area between alleys, usually tree planted, and could include all manner of contrived detail, such as significantly winding or irregular paths in the later gardens.

Thus the requirements for different functions were accommodated within the composition but they were always subservient to the principal lines of axes and their disposition within the whole could be quite arbitrary.

In Britain the avenues and hedges were usually of beech, a species that grows tall and straight, majestic in old age, and is ideally suited to the creation of garden artistry on an awesome scale. Pre-1962 photographs of Bramham show exactly what Benson had in his mind's eye 250 years earlier. Most of the other great plantings in Britain never got close to maturity before being swept away as victims of fashion. Beech hedges of course have the distinct design advantage of retaining their coppery foliage well into winter, so extending the period of effective spatial division without recourse to slower growing evergreens such as yew.

Flowers had a nominal or insignificant role to play in these garden designs, which probably contributed to the style never putting down deep roots on this side of the Channel. Britons, especially the English, had become a nation of gardeners in Tudor times, when they nourished a deep fondness for gillyflowers (carnations). These layouts had to appeal to a different set of values altogether and to people who were ready to embrace the concept of excluding some of the treasures of nature in order to appreciate the art of manipulating landscape elements on a massive scale. Grand Manner designs were intellectual, not comfortable, colourful or cosy. That said, whilst Le Nôtre felt that flowers disturbed the classical structure of his designs, not all his clients agreed. The consequence was that flowers were indeed used extensively throughout the seventeenth century in France.

Bramham Park

Robert Benson, John Wood the Elder, George London (probably)
From 1690

Bramham is a magnificent French-style park which miraculously survived virtually unscathed through the period of eighteenth-century 'Improvement' and beyond. The fact that the park has been maintained continuously despite the house having been ravaged by fire and lain unoccupied for the seventy-eight years following 1828 is a testament to the care which the Lane Fox family have bestowed upon their unique estate. In 1962 gales wrecked much of the woodland framework by removing over four hundred mature trees, mostly beech. This was followed by a replanting programme, the fruits of which are only now capable of being appreciated. It will be many years yet before Bramham's woods dominate in the way that they once did, but the structure has returned and with it the spatial subdivision which makes this one of the finest parks of any era.

Although heavily influenced by Le Nôtre, Bramham has some peculiarly English characteristics. The first of these is that the house is not placed centrally within the programme of the site layout. The French tradition was developed around the axis of the garden being inextricably linked to that of the house (Versailles, Vaux le Vicomte, Sceaux), leading to symmetry, duality and a degree of predictability. Only the parterre at Bramham, now replaced with an uninspiring rose garden, is related to the axis of the house. Vistas lead off in several directions inviting exploration beyond the immediate curtilage of the south-west front, and to follow them leads to the unexpected. Bramham has perhaps more of a woodland character than most French examples and relies on the carefully arranged avenues cut through the woods and wilderness plantings to relate the different parts of the plan to the whole. The visual link with the house is retained along only one of these avenues, the widest, which culminates in the T pond. The walk from the house is along a gentle incline which completely conceals the presence of this water feature until literally only a few metres away from it. To come upon it for the first time is to be truly surprised. The rigid framework of the avenue directs the movement, and the gradient of the land only reveals the full view as a reward for venturing so far from the house.

Some reference works attribute to Bridgeman the invention of the ha-ha. Others suggest it was first used by him as a garden device at Stowe, but a magnificent example, of the stone retaining wall variety, surrounds Bramham, allowing the adjacent fields to be included within the visual realm of the estate. That a French-style park should be enclosed so rigidly by a feature profoundly associated with the English Landscape Garden is a curious paradox, but the combination certainly works well. D'Argenville actually describes the Ah-Ah in 1709 as a component of the French gardens, so perhaps Stowe (from 1710) doesn't represent the revolution that it's supposed to. Indeed, there is also a fine ha-ha at Levens Hall from about 1690.

The composition at Bramham is both subtle and simple, consisting of grassed walks, trees and hedges (both predominantly beech), ponds, landform and garden buildings. The buildings are placed as focal points within vistas created by the high hedging and avenue planting. The vistas link the main gardens with the Pleasure Grounds of Black Fen, which are half as big again as the gardens. The physical link is by way of a series of pools stepping down into the steep valley which separates the two parts of the estate. These 'Obelisk ponds' are organised by stonework terraces, balustrades and cascades but they remain an unsatisfying feature of the design. The main water body here is broad, bland and sits in open space on the edge of the woods. The others are small and dank and are aligned in a south-east direction hemmed in by tall trees. Both scale and orientation work to the detriment of achieving success in this case as cascades are usually employed to best effect in full sun. The absence of enclosure at the main pond is in stark contrast to the rest of the estate and does not feel a comfortable arrangement. However, in the context of the whole estate this is a minor blemish which doesn't reduce the overall impact of Bramham's quality.

Quality is achieved through having a limited palette of material and a very strong masterplan with an easily recognisable relationship between each discrete part and the whole. These are important and guiding principles in the work of landscape architecture and ones which rarely fail to produce satisfactory results.

Bramham, despite its lack of variety, is a place for endless adventure. It is I believe the ultimate children's paradise. It has drama, beauty, contrast of scale from intimate and secret enclosure to the infinity of distant vistas (the balance between concealment and revelation), surprise and delight. The imagination can truly be set free here and there's not a Big Thunder Mountain in sight. It is large enough to get lost in, but at each avenue intersection there is the reassurance of a visual link to a part of the garden either glimpsed before or recognisable from the plan.

MELBOURNE HALL

Henry Wise
From 1699

Much truer to Le Nôtre's concept of site layout, albeit on a considerably smaller scale than either the Île de France estates or other examples in Britain, is the Marquess of Lothian's property at Melbourne in Derbyshire, an estate of only sixteen acres. Wise supplied the design 'to suit with Versailles', based upon drawings that his partner George London had acquired from Le Nôtre. Perversely, the principal axis doesn't align with the centre of the house but is offset to one side. Straightaway this lends a curious tone to the garden. During three hundred years of maturing it has undergone simplifications of detail, probably for ease of maintenance, and many of the hard edges have been knocked off. It is not therefore the ideal place to study the considerable design effects characteristic of French Renaissance gardens. However, what the design student visiting Melbourne can take away is the realisation that management of a landscape can completely change its character. Whatever the layout, change of use and change of resources will result in the designer's master concept itself being modified or lost. Although inconsistent with creating a design masterpiece, in an age of rapidly changing values it is as well to build in flexibility to landscape designs. Just as architects have found that those buildings which are capable of easy refitting can save their clients small fortunes, so the multifunction space as part of the simple design palette of open and enclosed areas can play an important role in the landscape designs of today. It's unlikely that Wise prepared a long-term management plan but at least the structural simplicity of his layout survives.

The main part of the garden is now open grassed terraces which offer very little of interest or inspiration. What makes Melbourne so special is not the interrelationship of the geometrically regular lines of the site plan, nor the ancient yew tunnel, nor even the 'Birdcage' arbour sitting at the end of the grand, largely sterile open space of the main vista from the house. The magic lies partly in the overgrown character furthest away from the house, which suits the relatively small scale of the garden. Mainly, though, this is a garden which either thrills or merely pleases depending upon mood, through all the senses. Take off your shoes and walk barefoot down the avenues in the southern corner, feeling the soft texture of spongy grass invaded by moss, being careful to avoid the wandering Muscovy ducks. Enjoy the intensity of temperature change between cool, dark green shade and bright, sheltered, sun-trap clearings. Listen to the cuckoos, woodpeckers and gentle fountains. Most importantly go in spring and take in great lungfuls of the pungent wild garlic – unseen but pervasive. You'll never forget it. At best, Melbourne can be such a tranquil garden that it seems to reassure that all is well with the world. Such well-being is not evoked easily. It takes considerable skill to create the conditions for fantasies like this to flourish and it's the management regime at Melbourne which has succeeded in this endeavour where many others have failed.

BOUGHTON PARK

Leonard Van de Meulen, (possibly Bridgeman), the Dukes of Buccleugh
Early eighteenth century

The estate at Boughton House is enormous and has been the home of the Dukes of Buccleugh since the house was built there in the seventeenth century. The landscape design, originally by Van de Meulen, who, like William and Mary, came here from Holland, is distinctively different in scale and detail from any other Grand Manner design in Britain. The layout character is French not Dutch and was inspired by sites like Versailles, knowledge of which was gained when the 1st Duke was ambassador in Paris during the late 1660s. Van de Meulen's parterres were short lived, and in the 1720s the 2nd Duke employed Bridgeman, who may have been responsible for much of what is seen today, including over twenty miles of avenues. The most impressive single feature is the square 'Broad Water' which extends in canal form both north and south to other areas of the estate. An eighteenth-century turf mount, or mound, also square and truncated with a broad flat top, stands next to the Broad Water and is now echoed by a new land sculpture, 'Orpheus', which exactly inverts its proportions and scale with a reflecting pool at the centre. The 10th Duke commissioned this work by Kim Wilkie, which owes more than a nod to both Bridgeman and Charles Jencks, as part of the extensive restoration programme which started in the 1970s and is still ongoing. The grid on which so much of the plan is based makes this a unique interpretation of the French garden and one which relies heavily upon lawn care to maintain the clean and even gradients of its embankments. Like Bramham, the designer's palette is simple and has been expertly used.

There is an important collection of follies and eyecatchers in the park.

ST JAMES'S PARK

André and Gabriel Mollet
From 1660

André Mollet was largely responsible for bringing the French Renaissance style to Britain, first in the 1620s when Charles I summoned him. Travelling back and forth between Sweden and England, Mollet was eventually appointed Royal Gardener to Charles II specifically to design and build St James's Park. His nephew Gabriel joined him in this enterprise and André's book *The Pleasure Garden*, containing his ideas and design strategies, was published here in 1670, having first appeared in Sweden in 1651.

Charles II had been in France during his exile and had absorbed French taste so it was not surprising that he wanted the design to be in the Grand Manner, in so far as it could be without being able to centre the plan on the palace itself. Because of this background, the site is included here, although of course the park has undergone several remodellings in the intervening centuries and no longer resembles a Grand Manner scheme in any way. Furthermore, land was lost from the eastern end to create Horse Guards Parade and later, Buckingham Palace was built at the western end. The design was very simple, consisting of double avenues aligned with the boundaries as well as either side of a canal 775 metres long through the centre of the park on its long axis. The St James's Park we have today probably caters for the particular leisure demands of the twenty-first century better than Mollet's scheme would have done.

OTHER SITES

Other estates that display similar effects and which were created in the same era can be classified as being in the Forest Style. This nomenclature derives from Stephen Switzer, an author who sought to free up and simplify the London and Wise approach by advocating the planting of extensive woodlands which had rides carved through them. These estates were designed not for court entertainment but for private pleasures. St Paul's Walden Bury and Cirencester Park both display Switzer's desired characteristics and have very long avenues (several miles long at Cirencester) which extend beyond the wooded areas.

In the Grand Manner, one of the latest sites to be restored to its former glory, in this case after fifty years of neglect, is Wrest Park in Bedfordshire. English Heritage's twenty-year programme is selective in that elements from each period of the park's changing design will be saved. Thus, the original early eighteenth-century layout, probably by London and Wise, provides the axis of the Broad Walk, the Long Water canal and Thomas Archer's pavilion; a seventeenth-century French-inspired parterre garden made in the nineteenth century sits behind the house whilst a rose garden designed in the twentieth century is to one side of the axis; a Bath House designed as a picturesque ruin is close to one end of a sinuous water body which Brown converted from geometric canals in the mid-eighteenth century; a Chinese pavilion and bridge from the same time provide interest at the other end of Brown's canal; and an American Garden, perversely framed and formed with Portugal laurels, rhododendrons and yews, also vies for attention. When restoration is complete, this garden will be a patchwork demonstrating many different design approaches, though most certainly none in their purest form.

Hyde Park, in London, was a royal park before the public were allowed free access in 1637. As a city park it is very large, being over 253 hectares when Kensington Gardens are included. The two areas are contiguous and, although technically separate, they read as one. In development, there have been several phases, starting in the sixteenth century when the land was fenced and maintained as a hunting ground for Henry VIII and Elizabeth I, having been seized by Henry in 1536. Charles I had The Ring laid out, a track for carriage driving that meanders around the peripheral reaches of the park. William III built the processional drive called Rotten Row (originally *route de roi*) from Kensington Palace towards Westminster, an alignment which parallels Kensington Gore just outside the park. It wasn't until the early eighteenth century when, under commission from Queen Caroline, Bridgeman dammed the Westbourne stream to form the Serpentine. Later, the Long Water was developed within Kensington Gardens, upstream from the Serpentine bridge. Bridgeman also constructed a ha-ha to separate Kensington Gardens from Hyde Park. His work was completed in 1733. A spider's web of paths, mostly long and straight and occasionally marked by avenues, overlays what appears to be a rather random arrangement of open spaces and tree-planted areas. Some large open areas have found a new use in the last fifty years or so, as a venue for open-air concerts.

Drumlanrig Castle, Dumfries and Galloway, is a four-square mansion house enclosing a courtyard and dates from the late seventeenth century. There is an avenue arranged axially with the house, which is built on a raised platform to emphasise importance. Parterres to the sides and rear were laid out on level ground with embankments separating them as the gradient falls away from the house. This landform provides good viewing opportunities and is as bold as any in the country. The gardens were restored in the nineteenth century after their having fallen out of fashion for a hundred years. Now, the detail as recorded by Inigo Triggs in 1900 has changed somewhat. It is simpler and like a ghost of its former self.

The Dutch influence and gardens in transition

The late seventeenth century presents a confusing picture of contrasting design aesthetics being promoted simultaneously. The Tudor style of extravagant display within compartments was being edged out by the new French school, or 'The London Mode' as some contemporary sources put it. The order, geometry, symmetry and focus of large-scale masterplans, as recorded in the engravings of Kip and Knyff, appealed to landowners throughout Britain but the smaller scale Tudor, Jacobean and Caroline style gardens were still widely admired although many had been destroyed during the Commonwealth period. Significantly, the gardens at Hampton Court Palace survived because Cromwell himself lived there. From 1689, just as the Grand Manner was being uniquely interpreted for British landscapes and attitudes, the Dutch version of Le Nôtre's methods and design philosophy began to find favour following the accession of William and Mary. This typically involved what most historians believe to be the cramming of too much display and decoration into a French-style masterplan, always with tulips and topiary. The result might be seen as a kind of hybrid between the two established styles at the time and was considered by some to be an absurd arrangement. The most charming extant example is probably that at Levens Hall in Cumbria, which lacks axial symmetry but remains a delightful, if overgrown, topiary-dominated jewel of a garden. Westbury Court, Gloucestershire, is often cited as the finest example of the Dutch tradition in Britain. From 1971 its garden has been meticulously restored by the National Trust, but crucially the relationship between house and garden has been lost as none of the three houses to occupy this site has survived.

William's House of Orange lineage was celebrated in many gardens of the period by the lining out of orange trees in tubs and, as both William and Mary were keen horticulturists, whatever they approved of, others took note and followed. Their palace was Hampton Court, where evidence of the Dutch style persists.

At the turn of the eighteenth century the British favoured understatement within their grand schemes and at least the English landscape was accommodating of orderly, if not always completely logical, design. As described, both the Bramham Park and Melbourne Hall layouts display order but don't follow the d'Argenville prescription of symmetry and axiality with the house. The rigorous mathematical and geometrical planning of Versailles and Vaux elevates those gardens to works of art, as supreme examples of the type, whereas even the lost Grand Manner gardens of Britain were adaptations subjected to peculiarly local influences. Parts of the original plan for Chatsworth in Derbyshire survive, including the main axis centred on the south front. Although the detail has been somewhat modified in the intervening years, somehow this symmetry escaped the improving hand of Brown later in the eighteenth century. Alterations followed in the nineteenth century and the whole composition now resembles more closely a theme park than either a work of art or a place for quiet repose. Other estates such as Castle Howard, Stowe and Blenheim that all began life under French rule, as it were, will be discussed later as their chief importance in the history of landscape design development relates to the changes initiated in the mid- and late eighteenth century. This is the dilemma that confronts all histories of garden design: gardens grow and change over time. It might reasonably be imagined that because of their geometric nature, Grand Manner designs, of all the styles that history has bequeathed, are perhaps the simplest to restore to original intentions, but as such can be accused of being stuck in their own time warp, like living fossils. The search for purity of design is therefore usually a forlorn exercise and, as with the examples of Bramham and Melbourne, changes have often been absorbed which have arguably resulted in improvements.

Legacy

Country estates were sometimes hundreds of acres in extent, no part of which could be undesigned or left to chance. It follows that the process of laying them out was both time consuming and expensive. The levelling of ground, the precision required in the setting out of avenues, the digging of canals, the planting of trees – not randomly but in regular grids or bosquets – must have seemed like a wholly unrewarding effort. To achieve the design intention would take several generations, so these could be described as unreal gardens. They existed only in the minds of their creators, and barely two generations later their descendants were already embarking upon 'improvements' in the new fashion. If they were expensive to make, they also demanded the highest standard of maintenance: clipping, trimming, pruning, weeding, raking. Everything had to look controlled and in perfect order. Ultimately *les jardins français* were only really fit for royal ceremony and elaborate parties and then only for the society into which they were pitched. Although history generally points to other influences, it's easy to imagine a financial imperative for change and what may have been a rush to embrace a fashion which removed crippling maintenance costs at a stroke.

The main lasting change that this period introduced was that the house (or palace) and garden (or estate) were for the first time considered as a unified design. The best examples were well balanced and proportioned so that the treatment of outside spaces was given equal weight to those inside, with a recognition of the relationship between the two. Avenues were used extensively and we still use them when wanting to emphasise direction or build a sense of expectation, although now rarely within gardens, the scale of which has reduced dramatically in the intervening centuries. Of course, fountains have also survived, as have reflecting ponds. Again the scale and geometry of water features has changed but the late seventeenth century was when their widespread inclusion in British gardens was first employed.

Also for the first time, designing and making gardens was established as a profession of sorts. The French had education standards for anyone wanting to practise in this field. The list of studies is quite impressive: The student should be able to read and write. He should also know arithmetic and be able to draw – which referred to a capability in tracing designs onto the ground. Although there was no requirement to study sculpture or painting, geometry was essential to enable the measuring of paths, beds and so on, as was architecture, a skill needed for the design of terraces, balustrades and garden houses. There were other requirements associated with horticulture, soil science and weather prediction. Of all these hoops that a gardener had to jump through, mathematics was supreme. People thought that its study could provide the key to understanding ideas. Newly available were logarithms and calculus, the application of which assisted the construction of waterworks and terraces with great accuracy, and Le Nôtre used his knowledge of these studies to create optical illusions in his designs.

Of all the features described by d'Argenville and employed in every Grand Manner design, parterres have not survived as desirable garden constructs, despite a late nineteenth-century revival in the classical schemes of architects such as Blomfield. Any analysis of the parterre must accept that its relevance is limited to static, unchanging display. Intricate and convoluted patterns which require a labour-intensive clipping regime and which occupy space that cannot be used – don't walk on the carpets – have no long-term future, except perhaps in the context of a work of art. Even then, the word 'craft' might be a more accurate description. The chances are high that any parterre in Britain will be a reconstruction or revival created in the interests of re-establishing a degree of authenticity to a historic garden.

Throughout d'Argenville, as translated by James, there is mention of improving the natural advantages and redressing the imperfections and inequalities of the ground. In other words, only when the site has been levelled can it be considered perfect and suitable for making a garden. This mindset would be almost completely turned on its head by the eighteenth-century improvers, who couldn't disagree more with the idea of several acres of flat ground next to the house being used for parterres, each offering 'beauty and splendour wherewith it constantly entertains the eye when seen from every window of the house'.

The strict rules that governed the layout and detail of Grand Manner gardens essentially provided a pattern book for gardens in emulation of the architecture volumes of Vitruvius and Palladio. It's to the credit of most designs in Britain that the rules were interpreted rather than slavishly copied as this resulted in less predictability and certainly less sterility of layout.

Away from the details, large-scale gardens such as were created in the late seventeenth century provided the opportunity to those walking through them for different areas in turn to be revealed and concealed, much more so than the cellular arrangements of knot gardens and mazes of Tudor times. Spatial diversity and sequential or serial vision, with all the richness of experience that they bring to the table, became part of the designer's toolbox and have embedded themselves in landscape architecture theory to this day.

Design styles don't change overnight and there is no absolute date when it could be said that the French and Dutch influences were no longer considered appropriate and the revolutionary English style – in many ways the antithesis of the Grand Manner – had taken over. Indeed, it isn't even as simple as identifying (as I have just done) a country with a style, as the process of change from the Le Nôtre way to what the French called *les jardins anglais* was playing out concurrently in both countries. Queen Anne, who appointed Henry Wise as Royal Gardener, died in 1714, five years before Burlington's first Grand Tour, which seemed to give impetus to the earlier work of Bridgeman and Vanbrugh at Stowe and Castle Howard. It happens that the English passion for change and their widespread implementation of the new style was conducted with considerably more confidence and vigour than their nearest European neighbours could manage. For simplicity's sake, however, we think of the French influence fading in the 1720s. Straight lines start to give way to curvilinear ones and art imitates nature rather than subjugating it.

The garden layout at Melbourne Hall relies heavily on walling and evergreen hedging to define the main areas and circulation routes. Everything about the Grand Manner style was on a large scale yet this mini version by Henry Wise, now somewhat soft edged, is more rewarding than many a larger garden planned strictly along textbook lines.

3

THE ROMANTIC REVOLUTION
1730s–1820s

Principal designers/ practitioners	Good extant examples	Principal components
Charles Bridgeman	*Castle Howard, Yorkshire*	*Irregular, curvilinear layouts*
William Kent	*Stowe, Buckinghamshire*	*Naturalistic forms*
Lancelot 'Capability' Brown	*Rousham, Oxfordshire*	*Water in the middle distance*
Humphry Repton	*Blenheim, Oxfordshire*	*Framed views*
Sir John Vanbrugh	*Bowood, Wiltshire*	*Ha-has*
	Claremont, Surrey	*Concealed walled kitchen gardens*
	Studley Royal, Yorkshire	
	Painshill, Surrey	
	Stourhead, Wiltshire	

Everything changed in the mid-eighteenth century. The French and Dutch influences, themselves based on the Italian predecessor, were to make way for a brand new way of designing gardens. Logic, geometry and mathematics had been the basis of the Renaissance layouts but the new revolution that started in England, tentatively at first then with real gusto with William Kent's essays at Claremont, Stowe and Rousham, owed much to classical allusion, poetry, idealism and a fondness for the landscape painters of northern Italy. In short, the English Landscape Gardening Movement was the antithesis of everything that Le Nôtre had stood for. Early experiments had been made by Vanbrugh and Bridgeman, even as some of the most accomplished Grand Manner designs were being laid out in the first two decades of the century. Although Kent raised the bar it was his student at Stowe, Lancelot Brown, who, from 1751, stepped up and perfected a formula that every self-respecting landowner in the country wanted a part of. Palladian mansions located at the focus of Brownian landscapes became the height of fashionable taste and the pattern spread very quickly throughout the country. The prospective client base was much larger than could be handled by just one practitioner so a plethora of designers dissected the Brown formula and provided consultancy services widely, in emulation of him. The quality of these cloned layouts was rarely more than satisfactory, but the overall effect was at least similar. As his style spread north into Scotland it became clear that it was less easily adaptable to the more dramatic landscape of the Highlands. Of the major designers, only Repton worked in Scotland, at Valleyfield in Fife.

William Kent joined Lord Burlington on two Grand Tours. The Grand Tour was a journey, often quite arduous, conducted by men of means and taste as a kind of continuing professional development of its day. Hadrian's Villa was the ultimate destination for most grand tourists, the antiquities of ancient Rome being the chief source of inspiration. In Burlington's case it was also where he acquired statuary for his own garden at Chiswick, in the layout of which he was assisted by Kent. Chiswick House, a classic Palladian villa, survives intact as a classical example of its type, but the garden was always a work in progress and a place to experiment with the new methods of a more relaxed design approach. Notwithstanding this, there are two pattes d'oie (each arm of which terminates in an obelisk, an eyecatcher, a bridge or some other focus), a central axis that runs through the house with avenues to front and back, and a grove of trees laid to a grid in the Grand Manner. The layout in part resembles the poet Alexander Pope's own garden at Twickenham in that rides are cut through wooded areas which lead off from the central axis and the centrepiece is a wide lawn originally intended for bowling. Both designs now appear to have been hybrids caught between the purity of the styles which pre- and post-date them. Painswick, a garden started in 1738 at Stroud in Gloucestershire, is another site having some of the qualities of both periods. Allegedly the best example of a rococo garden in Britain, it has been the subject of a restoration programme since 1983. Whether or not 'rococo' defines a distinct style in garden design is a moot point, but in essence its chief characteristic is eclecticism.

For those designers working without any great creativity of their own, there were practical reference books which guided their hands. Philip Miller's *The Gardener's Dictionary* of 1731 concentrated on matters of a horticultural nature, while John James' translation in 1712 of d'Argenville, under the title *The Theory and Practice of Gardening*, as mentioned earlier, provided advice on design, albeit somewhat outdated. Indeed Miller borrowed large elements of James' book, specifically on the techniques of laying out gardens, and reproduced them in his. Several poets and essayists, particularly Joseph Addison and Alexander Pope, described a new, more relaxed way to lay out gardens which reflected the inherent qualities of the landscape rather than fighting against them. Indeed they ridiculed the Grand Manner and all its conceits, like topiary and parterres. Pope made his own garden which proved to be a significant inspiration to others. The cognoscenti, led by Burlington, were influenced by philosophical writers, by Arcadian landscape paintings mostly of

Italian scenes and by their own travels on the continent. The artists Claude, Poussin and Salvator Rosa were much admired for their landscapes of the hills around Rome, depicting temples and ruins as focal points.

The Sublime and the Beautiful

Edmund Burke's essay *Philosophical Inquiry into the Origin of our Ideas on the Sublime and the Beautiful*, of 1757, explored and defined various human emotions, a study which contributed to the wider endeavours of the Enlightenment. Burke identified man's chief preoccupations as procreation and self-preservation, resulting in a keen awareness of objects, scenes and events that stimulate the emotions associated with these activities. Whereas thoughts of procreation will be stimulated by the curves of a woman's body, it follows that anything softly curved or undulating can be associated with beauty. Hogarth's *Analysis of Beauty* reinforced Burke's views and coined the term 'the line of beauty'. Brown's lawns, lakes and winding driveways fell into the classification of beautiful. By contrast, anything that was very large, such as mountains, valleys, cliffs, waterfalls and gothic ruins, or which was menacing by its presence, like wolves, bandits or thunder, stimulated a sense of endangered self-preservation. The objects that brought out these thoughts as well as the sensations themselves were dubbed sublime. No single garden can be considered to be completely sublime, although aspects of Castle Howard certainly qualify. Most eighteenth-century gardens had areas devoted to the creation of sublime effects. The north end of the lake at Stourhead, planted with pines, was one example. Strangely perhaps, some designers, William Kent included, 'planted' dead trees for sublime effect.

Some 250 years later both the Sublime and the Beautiful are qualities that are still admired in garden aesthetics. Distant views of mountains or the sea nearly always command a premium, as do smooth and extensive lawns leading down to serpentining lakes.

The Picturesque

The term 'the Picturesque' denotes an influence which affected the final years of the Landscape Gardening Movement, from approximately 1790, although its origins were earlier in the eighteenth century. William Gilpin introduced the concept and gave it a name in essays published from 1782. Broadly speaking, Picturesque gardens employed the same techniques as the landscape painters to create landscapes in three dimensions in imitation of those that had been made in two. It was seen as an approach to landscape design that allowed both the Sublime and the Beautiful to coexist. Picturesque gardens were more complex than Brown's designs and were a reaction against their simplicity. Variety and contrast were deemed necessary as a replacement for the peace and tranquillity of middle-century layouts. Again, parts of Stourhead illustrate Picturesque principles, as does the last stage of development at Studley Royal, which had its own ready-made ruins in Fountains

Abbey. Painshill is one site among many where artificial ruins were constructed for picturesque effect.

Richard Payne Knight summarised the difference between the Beautiful and the Picturesque. The former he likened to 'the garden dressed in the modern style' and the latter to 'the garden undressed'. Picturesque designs required only nominal design work to bring out their essential qualities, which were essentially wild in character.

Humphry Repton took the theories of Payne Knight and another writer of the period, Uvedale Price, and applied them to landscape design. The result was that, just like paintings, there was a foreground, middle ground and background, each having a different character. He believed that art should dominate the foreground, which translated into geometric beds and colourful planting, the middle zone should be parkland in the style created by Brown, and furthest from the house the landscape should be of a wild character, or at least 'natural' in appearance.

Design detail

Charles Bridgeman is generally considered to be the designer who made the initial break from the Grand Manner. The changes he implemented were modest compared with those that followed but included frequent use of the ha-ha and some irregular path layouts. The period of his professional career spanned both eras and because his apprenticeship was served under London and Wise it is only to be expected that geometry always dominated his layouts. His own acceptance of change was a slow one, as was that of the introduction of the landscape gardening approach throughout the country. As Wise's career was ending, Bridgeman taking over as the new Royal Gardener, there was increasing concern about the cost of maintaining the royal gardens. Together they prepared a report, in 1727, which suggested simplifying the layouts to save money. This may have been the ultimate catalyst for change, whilst poets and writers also provided various rationales in favour of naturalistic designs.

William Kent, working usually on a relatively small scale, popularised the concept of a planned circuitous route through the garden. This enabled a controlled sequence of views to be established with contrasts of light and shade, open and enclosed spaces, concealment and revelation. The garden became the canvas upon which to create a collection of tableaux which the visitor discovered as a variety of little adventures and the house was almost irrelevant to the gardens. The Elysian Fields at Stowe are a completely self-contained design within the larger masterplan of the estate. Equally, the part of Rousham which he redeveloped from the Bridgeman original is quite removed from the house, visually and physically. Kent had his detractors. Horace Walpole disliked his layout at Holkham Hall, Norfolk, referring to the tree planting there: 'Mr. Kent's passion, clumps – that is sticking a dozen trees here and there until a lawn looks like the ten of spades' (quoted in W. O. Hassall's article in *Garden History*, spring 1978). By way of contrast to Kent,

Brown allowed the house to take its place in the centre of the design, not at the end of a grand avenue but in a prominent, open location with grass sweeping right up to the garden front. From the time when Brown set up his own business in 1751, his stripped-down formula dominated as the preferred landscape design aesthetic. Apart from locating the house to best effect, recognising the genius of the place, the other main features of his approach were:

- A surrounding belt of trees for the entire estate. He used only seven species of tree on a regular basis.

- Clumps of trees set within open parkland and on the inner side of the surrounding belt. This was an idea borrowed from both Southcote and Kent.

- A water body (lake or widened river) in the middle distance, often linked to a cascade. Where the water flowed in and out was always concealed from the view from the house and the banks of the lake or river were always grassed. Marginal planting would have obscured the serpentine effect of the line of beauty.

- No visible outbuildings were allowed. The kitchen garden was walled and remote from the house. Tunnels were also employed to conceal the comings and goings of servants between the house and their own quarters nearby.

- Ha-has were often used to give the impression of the house standing in the vastness of the wider landscape, whether cultivated or wild.

- Apart from the big house, the landscape dominated Brown's compositions. He only reluctantly used garden buildings or ruins within his compositions, which were essentially green, with little attempt at colour or textural variation. Indigenous trees such as beech, lime, elm, oak and Scots pine were favoured, although Lebanon cedars were also a favourite.

- A curving approach drive rather than a triumphal avenue.

This formula, although still admired, has also been much criticised, both in Brown's own time and ever since, mostly for its blandness and lack of originality. In 1978 the Victoria and Albert Museum staged a Garden Exhibition in which none of Brown's work was illustrated but where one caption read:

Retreat of the garden: Capability Brown was the arch destroyer of the garden as a formal and informal element near the house. He was a landscapist rather than a gardener and the fashion for his natural style of parkscape began to grow from about 1750. His practice was built upon important precedents: William Kent's arcadian garden style of the 1730s; a proto-naturalism due to the damming of lakes, and the simplification of complicated formal gardens around the great houses due to economic pressures. Brown's disinterest in horticulture denies him a place in this exhibition.

The Picturesque Movement embraced garden buildings and ruins, often mock-ruins, as components in the layout of the estates and they helped to provide a focus that was in some cases previously lacking.

The early terraces

In north-east Yorkshire there are three similar examples of the early breakaway from the Grand Manner designs. Chronologically, the terrace at Duncombe Park came first, having been laid out between 1713 and 1730. It's very likely that Bridgeman was responsible for the earthworks which levelled the ground on the east side behind the Great House and from where a long view was opened up over the valley of the River Rye. Vanbrugh designed one of two classical temples which mark the curving terrace's end points. In similar vein, the Rievaulx Terrace, a few miles distant but still within the Duncombe estate, was constructed on high ground overlooking the ruins of the abbey, probably in 1758. Here there is more emphasis on picturesque effect, a great effort having been devoted to giving the impression that the landscape is a natural one. Visitors to the terrace today will perhaps quite readily recognise the artifice. On the wooded hillside below and between the terrace and the abbey, twelve narrow viewing corridors have been left clear, each focusing on the ruins below but of course from a slightly different perspective. Temples again mark the end points, half a mile apart, and tree planting on the east side of the terrace helps to direct the view to the west. In this respect the terraces are very similar because there is a strong element of control in the planning of both. Spatially, the Duncombe Terrace has more variety and perhaps also more lasting appeal in that the off-site focus is less obvious and repetitive. The third terrace is at Castle Howard and is detailed below.

The terraces demonstrated what was possible when the strictures of geometric planning were relaxed. Interestingly, all three terraces relied upon the wider perspective for their effectiveness, a device eschewed by Brown.

The ferme ornée

William Shenstone was an influential character in the early development of English romantic literature though he never attained the credit that he felt was due. He was born and lived at The Leasowes, an agricultural estate in Halesowen, and spent the last eighteen years of his life, from 1745, improving it in line with the contemporary fashion. This, not his poetry, was his legacy. Like Woburn Farm in Surrey before it, The Leasowes was a ferme ornée, a term coined by Stephen Switzer, literally an ornamental farm. Philip Southcote's Woburn Farm, dating from 1735 and where Kent was involved in designing certain of the architectural features, was allegedly an inspiration for Brown. The ferme ornée was perhaps a little closer intellectually to his approach than the intimate-scale theatrical compositions that Kent devised at Rousham and Stowe. The idea was to elevate the working pastoral and arable landscape onto a higher level, displaying man's total harmony with nature. This Arcadia was alluded to by locating classical statuary and texts within a framework of circuitous footpaths around the farm. Grottoes and buildings with no purpose other than their having picturesque qualities were sited to be viewed from the pathways to best effect. The paths, which were fenced off from the pasture, were also adorned with woody shrubs and herbaceous perennials. Other design features such as water bodies and framed views formed part of the design palette. The Leasowes, as a ferme ornée, has been lost and the site is now the Halesowen golf course, but Dudley Council is restoring the park as far as possible to its 1740s form.

It seems that the genesis of the ferme ornée was entirely English, despite its nomenclature suggesting a French origin. It was merely fashionable for the educated classes to use French whenever they could in an obvious display of their level of cultural sophistication. The ferme ornée was an upper class ornamented farm.

Hagley Park, in Worcestershire, is another site. It was designed for Lord Lyttleton 'after plans by William Kent' sometime after 1739. A path follows a circuitous route through the estate and follies are located at strategic positions for picturesque effect.

Southcote and Shenstone's attempts at fusing art and agriculture were, ultimately, nothing more than an interesting experiment indulged in by wealthy landowners. The idea that the working landscape could be aesthetically improved on anything other than an occasional small scale was probably never going to be realised. The ferme ornée remains a footnote in the history of landscape design, financial and practical common sense having intervened. They may have been lost many years ago but their legacy, pointing the way to what was to become the dominant style of the English Landscape Garden, is important and deserves proper recognition.

CASTLE HOWARD

Vanbrugh from 1709, Nesfield 1851

The ability of the device of the avenue to stir emotions is nowhere more powerful than at Castle Howard, created for the 3rd Earl of Carlisle. Unlike the grand avenues at Stowe and Blenheim, here it's not focused on the house but runs past it for five miles on its west side. The effect is similar from either direction but there is perhaps more drama created when turning onto the road at the north end from the winding narrow lanes of the gently rolling Hambleton and Howardian

Hills. The sudden change from a relatively small-scale domesticated landscape to a deliberately contrived statement of grandeur and importance with a double-row beech and lime avenue is quite enough to get the heart racing of anyone fortunate enough to experience it. The folds in the landscape are still there, as they are at Stowe, and this has the effect of heightening the sense of expectation as the long view is always changing until its climax, the obelisk, is at last revealed. The obelisk marks the entrance drive to the house with a 90° turn to the east and acts as a punctuation within the long avenue offering the opportunity to deviate from it and so gain access. The destination has already been partially viewed under the avenue's canopy, where the north front of the house is set beyond a foreground of the Great Lake. This is large-scale landscape planning employing the classic conceal and reveal technique to raise expectation whilst offering a glimpse of the main event.

Castle Howard was built on the site of an old village, Henderskelfe, the main street of which was grassed over and used by Vanbrugh as a gently curving terraced walk connecting the south front of the house with an entirely new landscape composition which he fashioned at the road's other end. As such, the walk, or terrace, qualifies along with the terraces at Duncombe and Rievaulx as one of the first essays in the new oeuvre of landscape design. Notwithstanding that Castle Howard was Vanbrugh's first architecture commission, albeit with the assistance of Nicholas Hawksmoor, the landscape he created here was also truly groundbreaking and demonstrated a clear understanding of some basic landscape design techniques. The Terrace Walk culminates at Vanbrugh's Temple of the Four Winds, a perfectly proportioned garden pavilion commanding a view beyond the estate grounds towards the Howard mausoleum, designed by Hawksmoor, on high ground to the south-east. A three-arched 'Roman' bridge in the middle distance spans an artificial river and a pyramid folly lies in fields to the south. All of this makes a picture which is best viewed from the temple, from where the techniques used are most easily revealed. The mausoleum is a domed building which stands on the crest of a rounded hill, emphasising the natural contours of the land. A woodland partly obscures the full view, inviting further exploration. The bridge and 'river' similarly draw attention to the lowest lying land, extending the visual depth of the scene and adding incident to an otherwise relatively dull landscape. Together they double the vertical range of the view and establish a dynamic previously absent. The bridge was an addition of the 1740s, after Vanbrugh's death, but whether he envisaged it as part of the composition is not really relevant. What is important is that here is a landscape design on a grand, heroic scale, the various parts of which relate to each other and to the whole, providing an early example of the style which would become the English Landscape Garden.

Carlisle had earlier called in George London to plan aspects of Wray Wood, which lies on high ground to the north-east of the house. His starburst design, common at the time, was rejected because Carlisle was interested in being creative.

William Nesfield completed the site layout by designing the south parterre, its centrepiece being the Atlas fountain from the Great Exhibition of 1851. Originally this was an extremely intricate design but it is now splendidly restrained and somewhat out of character with the prevailing Victorian custom. Its components are only yew hedging and grass, linked with gravel paths, consistent with those elsewhere in the gardens, to provide the essential continuity.

Rose gardens, a time capsule and a garden centre have all been added to the delights of the estate in modern times. See past these to the grand planning of the early eighteenth century and recognise a magnificent work of landscape design which, once experienced, will never be forgotten. It's an original work of art from one of the most inventive periods of design in the country's history. The best garden building anywhere in the country, the Temple of the Four Winds, is alone worth the visit.

CLAREMONT

Vanbrugh, Bridgeman, Kent, Brown
From 1715

By contrast with its close neighbour, Painshill Park, Claremont has seen many changes to its design over the centuries. All the great practitioners worked here, in turn contributing something of their own style to the garden but leaving it as a rather confused and unsatisfactory composition. The garden has always been visually separated from the house, as is Stourhead, the backbone being an existing ridge of high ground, the mount, separated from pastures on either side by ha-has. The politically settled period since the Restoration, which had given rise to a general expansion of building activity, suffered a blip in the year that garden work started at Claremont, with a Jacobite Rising in the Scottish Highlands. Thirty-one years later, when the more serious Rising was put down at Culloden, the garden was still being worked on by William Kent and was to see at least another twenty-five years of development.

Vanbrugh bought the site and built himself a modest house there in 1708 but later sold it to the Earl of Clare who commissioned Vanbrugh to design and build him a larger house, and from 1715 he started the process of making the landscape gardens. The Belvedere, a garden building by Vanbrugh, stands on the highest part of the mount and an avenue follows the line of the ridge down to a rectangular bowling green at the lower end. Bridgeman was involved with the first layout at Claremont and by 1725 the signature feature of the gardens, the great turfed amphitheatre, was completed to his designs adjoining a circular lake near the south-west corner of the site. It is an earthwork of impressive proportions (over three acres) and geometric regularity. Whether it was ever of great utility or convenience is another matter, but what Bridgeman created survived the reworking of the gardens by Kent in the 1730s and 1740s and even the wholesale changes implemented by Brown for Lord Clive from 1768. Brown planted cedars and other trees on the amphitheatre without first regrading the land, so when the National Trust came to restore the gardens in the 1970s they found that Bridgeman's terraces were largely intact beneath the later plantings. If Brown's work is hard to find here, the same is not true for that of Kent. He enlarged and reformed the margins of the circular lake, giving it a naturalistic character, and added an island and several garden buildings. A grotto on the lake's south-eastern bank was added at about the same time. The reworking of the lake set up an uneasy visual relationship with the amphitheatre – the clash of styles being clearly of less importance to Kent than his respect for Bridgeman's earlier work.

As alluded to already, landform tends to endure far longer than planting or even buildings, the earthworks at Claremont being amongst the most celebrated examples in British garden design and serving to underline that fact.

Claremont has characteristic design elements drawn from all the different twists and turns in design development during the eighteenth century. As such, it can be viewed more as a museum of the era, a collection of parts rather than a unified and coherent design. There is more for the curious student of garden history than for anyone seeking a classic example of the English Landscape Garden.

STUDLEY ROYAL

John and William Aislabie
From *c.*1720

No history or design exploration of the eighteenth century would be complete without including the park at Studley Royal. This water garden is the work of amateur designer and professional politician John Aislabie (he was Chancellor of the Exchequer) and later of his son William. Aislabie was of course in a good position to meet and be influenced by the best garden designers of the time, but the layout and detail at Studley Royal appears to have been entirely of his own conception.

The River Skell, in its narrow and wooded valley, is the central feature linking all other elements of the composition. Work progressed in an upstream direction starting with the creation of a lake which itself was fed by a canalised section of the river by way of a rather understated cascade. Either side of the canal Aislabie made geometric ponds, the circular Moon Pond being the main attraction and providing the foreground to a later garden building, the Temple of Piety. Turfed embankments recall the work of Bridgeman at Claremont, with which they must have been approximately contemporary. The work implemented by William involved widening the river upstream of the canal to form the Half Moon Pond and footpaths along the valley to Fountains Abbey. The Abbey ruins were a ready-made picturesque incident and needed only to be opened up to view. They were incorporated into the estate in 1768. Both father and son reflected the prevailing spirit of their age in the works that they undertook at Studley. John's improvements were mostly in an obviously contrived manner whereas William embraced the naturalistic and Picturesque principles that dominated the approach of mid- to late-century designs. The two styles blend seamlessly to make a garden experience which is inventive and unique. This has been achieved through the careful use of the existing landform and the planting, particularly of evergreen hedging, which together organise the space and control the movement through it.

Elsewhere in the estate there's a very impressive avenue leading from the entrance gate and aligned on the steeple of St Mary's Church, although originally an obelisk occupied the focus. Access to the water gardens, the seven bridges walk, the deer park and all other areas is gained from this avenue, itself a feature that betrays the era of the earliest design.

Attention should be drawn to an obvious design weakness within the composition. The path that runs on the outside bend of the Half Moon Pond is located not at the river's edge but many metres away at the foot of the densely planted river bluff. Walking on this path is a less than comfortable experience because the water's edge is a magnet denied by the site planning. People generally like to position themselves at spatial transitions, feeling less happy about being in the centre of wide open spaces. True, the path is located at an edge, between open grass and wooded bluff, but the water has a more powerful attraction, and always will do. At the running out of the bend, Fountains Abbey comes into view with the river and Abbey Green in the middle distance. The path moves closer to the river bank and is less of a site planning issue.

The path detail aside, the water gardens and the woods either side of them present a series of sequential experiences, tightly controlled and full of variety. Being a hybrid design, transitional between Baroque and Picturesque, places Studley Royal in the same category as Claremont, albeit with much more cohesion as a result of staying true to the masterplan. The Aislabies certainly knew how to get the most out of their site and their legacy is a remarkable one. Whatever the season or varying weather and light conditions, the water parterre zone is always a quiet, almost private space, proportioned and detailed to perfection.

PAINSHILL PARK

Hamilton
From 1738

This is a site that has been resurrected from oblivion since the mid-1970s and now represents one of the finest essays in landscape gardening anywhere in the country. Originally the layout had two distinct zones, only the earliest of which has been restored. Hamilton's later works, in the manner of a Brownian park, lie to the north and outwith the current ownership boundary. The garden is in The Picturesque style and, like Stourhead, is centred around a large lake of about thirty acres with islands. Dotted around this landscape are garden buildings and bridges, each carefully located for maximum picturesque effect. Fortunately the vast majority survived the period of neglect from 1948, each providing a focus for Hamilton's recreated vistas. A two-times Grand Tourist, Hamilton was a gifted amateur in the field of designing and detailing landscapes and he concentrated his efforts on this, his own property, assisting others with theirs only occasionally. More diverse in its detailing and mood creation than its nearest design counterpart Stourhead, Painshill is also a true masterpiece of its time. In both, the house is quite separate from the landscape garden. Painshill's plan still looks very modern, the disposition of open spaces to both wooded ones and open water being in the approximate proportion of equal thirds. This successful formula was part of the design strategy at the Amsterdamse Bos Park (begun in the 1930s) and has since become a recognised model worldwide. Interestingly, the general perception that people have of such places is that they are fairly heavily wooded. The River Mole, which forms the southern boundary of the park, makes several meanders which are echoed by the line of the lake just to the north. The two are separated by a narrow causeway, heavily planted with trees for visual enclosure. The lakeside footpath offers views of the vineyard, amphitheatre, bridges and buildings at intervals along its length, the scale varying constantly and thereby holding the interest of the viewer.

In a different approach from Hoare at Stourhead, Hamilton indulged his passion for exotic trees and shrubs, not as individual specimens but as part of the general composition of mass and void. If the plan looks modern, so too does the experience of the spatial sequence which, again in contrast to Stourhead, allows for options to be chosen. Studley Royal, for all its qualities, is chiefly a linear walk which must be retraced when the far end is reached. The terrace between the house and the Temple of the Four Winds at Castle Howard is the same. Kent and Brown introduced plans which, like Painshill, allowed more freedom to explore, a feature which was incorporated within the design of public open space from the earliest Victorian parks onwards.

Blenheim might have the most impressive lake and, for some, 'the finest view in England', but Painshill has a unity and unselfconscious beauty that it would be hard to imagine could be better contrived.

STOURHEAD

Hoare
From 1743

In many ways Stourhead is the quintessential early form of the Landscape Garden and remains one of the most loved and most visited in the country. It is the work of an amateur designer, the banker Henry Hoare, and was created over a period of thirty-three years from 1743. Damming the relatively steep-sided valley produced a roughly triangular lake. Together these two features provided the structure for the creation of the garden. Tree planting on a large scale clothed the valley sides and a circuitous footpath was threaded through these new plantations. The path controlled the views of the incidents on its route, such as temples, bridges and a grotto, all of which were carefully placed for maximum picturesque effect. In the design of the various buildings, Hoare received professional help from the architect Henry Flitcroft. The overall intention was to mimic the paintings of Claude Lorrain and Gaspard Dughet but in a way which brought them to life from the ever-changing vantage points around the lake.

The plan is very much in the Kentian mould of course, but the nineteenth century saw the introduction of numerous exotics at Stourhead, particularly massed rhododendrons, the effect of which was an inevitable character modification. As with many of Kent's gardens, the house is quite separate and doesn't interact with the garden by being a prominent or even visible feature within it. The pathway was always intended to be followed in an anticlockwise direction, in which way the views would open up in sequence as planned. This is where the chief quality of the design rests. Every opportunity is taken to vary the experience along the route, the path following the contours within dense planting, then grading down to the lake, up again and through a tunnel, suddenly an open area next to the lake and always framed views are presented, with water usually in the foreground. Manipulating the position of the viewer in this manner ensures that nothing is ever missed and that, in this case, the experience is a rich one.

STOWE

Bridgeman from 1713, Vanbrugh from 1715, Kent from 1735, Brown from 1740

Stowe rightly has an important place in the history of designed landscape in Britain. The layout of this very large estate has, like many others, been transformed over the centuries, but many of the basic components which comprise the essential programme are still there to be admired and studied. The Palladian mansion stands on high ground commanding the whole estate and lies on the central axis which passes through open lawns (parterres once covered the area which is now occupied in summer by a full-sized cricket pitch), the Octagon Pond, triumphal Corinthian Arch and a double avenue focused on Buckingham. The approach to the house from the south, so well outlined by Sylvia Crowe in *Garden Design*, is an object lesson in site planning and control of effects, involving two axes of the original Bridgeman design, and ultimately places the traveller at the south portico of the house looking back on the first part of the approach, with water in the middle distance and the Corinthian Arch skylined as a silhouette. Today, this axis is strongly controlled visually by mature trees and underplanting through which various paths lead to a circular route of the estate. The Corinthian Arch, containing two four-storey flats (originally for gamekeepers), was built in 1765, well after Brown had left Stowe, so the components of the now famous approach sequence evolved very slowly and are, like the gardens themselves, the work of many hands, not a single masterplanner.

On this scale, the designers have had to accept the existing landform, changing it only subtly to suit the masterplan. Contrived landform, as a space-making device, could only be successful on a much smaller scale in the days before mechanical scrapers and dozers, so the principal divisions of most layouts, as here, are created with trees. At Stowe an impressive earthwork surrounded the larger part of the estate but even this military-scale trench was not for visual enclosure, quite the opposite in fact. Bridgeman's ha-ha enabled the gardens to be surrounded by a secure but invisible boundary, now restored as part of the National Trust's comprehensive programme which was initiated in the 1980s. Modern use of ha-has includes that at Grey Walls (1900) where Lutyens included the 18th green of the Muirfield golf course within the visual realm of his garden design, which is otherwise subdivided by high walls. More recently the circle has been squared in animal collections throughout the world where a 'safe danger' design approach has largely eliminated the need for cages and high fences.

Stowe, which has been a public school since 1923, used to offer the best value of all destinations on the college grand tours, the obliging school authorities allowing educational visits without charge. Since the National Trust became custodians of the landscape here it is a different story, but the approach sequence is entirely within the public domain and can still be appreciated without having to part with cash. I suggest however that the climax of the sequence, where, at one glance, the logic of the design is exposed, is worth paying for. For many years yet the replanted avenue, which forms the first part of the approach, is going to be a visual disappointment, particularly for those who have experienced the avenues at Castle Howard. The trees are an emphasising feature of what is obviously an uncompromisingly direct and yet rhythmic axis. The landform controls the vistas within the avenue, in turn revealing and then concealing the distant house. The full effect can only really be appreciated by driving slowly. Just as the Corinthian Arch frames the distant house, the track turns off axis and rejoins a direct route to the north forecourt at a point where the two Boycott Pavilions stand either side of the road as another gateway to the estate. The entrance door of the house, on what is called the Park Front, leads to a square hall which itself leads to the imposing oval, colonnaded saloon, its doors aligned with the main axis of the whole estate and giving onto the south portico on the Garden Front. The ordering of the estate is at once revealed and a finer prospect is hard to imagine.

There are many garden buildings surviving from the eighteenth century although many more have been lost. Most are worth studying in their own right as examples of fine architecture as well as components in the designed landscape, where their relationship to the path and therefore to the eye of the viewer is handled with great skill. The wooded valley to the east of the garden axis contains the largest concentration of buildings and monuments, arranged either side of a stream. This is Kent's Elysian Fields (from about 1735), within which juxtapositions invariably involve the artificial stream as a physical barrier between buildings and a visual enhancement to the contrived picture, which is the reward for arrival. The landscape cannot be hurried. It is peaceful, well balanced and secluded and, like Rousham, is a garden within a garden, confirming Kent's preferred scale of working or at least that in which he was most successful. His sketching talent may have been rather limited but there is no doubting his ability to turn ideas into high-quality, romantically classical landscapes.

What is acknowledged to be Capability Brown's first essay in landscape design after moving from the Northumberland estates where he cut his teeth, the so-called 'Grecian Valley', is worth viewing if only for academic interest. Brown's forte lay not so much in modest tinkering with established layouts but in wholesale and grandiose change. Historians never accuse him of timidity in design but at Stowe he was still learning his craft and was apparently firmly under William Kent's control. The Grecian Valley is a curious, 'open bowl' kind of space surrounded by trees, the front line of which is broken to incorporate occasional temples. To wander through the space is to feel unquestionably isolated from the rest of the estate. There is nothing about it to suggest that the area forms a coherent link with the rest of the design. It is what it is, an afterthought. Designers should beware of falling into this trap when called on to modify or extend existing compositions. First, study the site; second, appraise the possibilities; third, advise the client on the merits of different solutions; and fourth, turn down the commission if the landscape is clearly going to be the loser. The Grecian Valley brings nothing to the bigger picture that is Stowe and, had it never been developed, it wouldn't have been a great tragedy. The experience it gave Brown is another matter and in the long run was of considerably greater value. Brown may also have assisted with the layout of Hawkwell Field to the east of the Elysian Fields. This was designed as a ferme ornée, where grazing cattle and sheep were allowed free roam, the area containing several buildings by James Gibbs.

I have a soft spot for Stowe, particularly the avenue from Buckingham. On one of our study tours the students had discovered that our visit to Stowe coincided with my birthday, so when a spontaneous chorus of *Happy Birthday to You* broke out in my minibus as we drove up the avenue, a little unnecessary tension was added to my personal experience of this landscape drama. I like to think of that moment as a sensitive tribute – that fifteen Year 1 students could already recognise a significant landscape feature and mark it in that way. On the other hand they might have been thinking about their end-of-year grades.

Rousham

Bridgeman from 1719, Kent from 1738

Rousham, pronounced 'Roosham' locally, is quintessential William Kent, although it must be stressed for the record that the main lines of the design are not his but Bridgeman's. As might be expected then, this landscape garden is rigidly structured by a geometry which relates the parts to the whole and determines the scale. It would be fair therefore to give more recognition than is evident in most accounts to the earlier layout in its determination of the later one.

This is a small garden by the measure of many of its contemporaries and is constrained by landform, a public road and a river. That it includes the wider landscape within its realm is characteristic of the English Landscape Garden. In this case the floodplain of the River Cherwell is the borrowed landscape and the river the separating feature. At Levens and at Bramham we find ha-has surrounding the gardens and yet at this, one of the first of the seminal designs of the new movement, the ha-ha is used to bring the 'natural' landscape right up to the house itself. It takes the form of a stone retaining wall and steeply sloping ditch separating the paddock from both the house and landscape garden. However, it is in the garden where the student will find most to admire and learn. We don't know whether Bridgeman intended it to be seen in a particular sequence, but Kent certainly did and it is his carefully contrived views and spatial juxtapositions of the 1738 plan which display the beginnings of the style which is still held to be the only original art form England has contributed to world culture. The sequence is well enough described in the estate's guidebook and in numerous history texts and is not particularly relevant here. Acknowledging that whilst a landscape architect will be interested in the detail of its layout, the garden gives the visitor more than a collection of fine buildings and spatial experiences. Just as at all really great gardens, there is an intangible element which derives from the total composition being right for the site. Everything about the garden at Rousham is perfectly balanced, both the independent effects and in each adjoining part's interrelationship. To follow Kent's intended route from the house and back again is to experience a delightful, surprising and peaceful garden full of variety and contrast. So often the most satisfying designs are created with limited space and a restricted palette. Rousham is one such example, designed on an intimate scale consistent with the lack of space available.

The footpath linking the individual features is barely one metre wide – insufficient to allow two people to pass – and yet if it were any wider there would be a jarring of scale. The spaces are today very much more clearly defined than Kent had wished and, judging from his sketches, they are better for that. Trees and undergrowth form complete visual and physical barriers, the foils for the artefacts and backdrops for the spaces. They determine the size of the outdoor rooms and corridors and they emphasise direction where appropriate.

Into this sylvan paradise it was once hard to ignore the influence of adjoining land uses. Whilst the visual realm extends across the valley of the Cherwell through which a road and railway both run, these are of no consequence at all compared with the erstwhile RAF base at nearby Upper Heyford, which closed in 1994 but had been the station for F100s and F111s of the United States Air Force Tactical Fighter Wing. There should be no place in a magical landscape such as this for machines of war with ground-trembling afterburners and, thankfully, they have moved on. The woodpeckers now dominate and have no competition to speak of.

Rousham is as poetic as gardens come. Not only does it represent the turning point in history when the French influence was largely discarded for a character more in tune with the native qualities of the English countryside, but it is preserved largely intact, retaining a spatial design which is at least as effective now as at any point in its past. Like Kent's Elysian Fields at Stowe, the landscape garden at Rousham is small in scale. The spatial design takes account of the river valley topography in allowing prospects to be opened up and then concealed. The 'eyecatcher' folly placed on the skyline to the north adds a picturesque detail to the scene viewed from the bowling green next to the house. This view literally looks over the landscape garden, which lies at a lower level, and confirms the notion that landscapes were designed as a series of carefully composed set pieces. Compare it with the view from the Temple of the Four Winds at Castle Howard, which has more drama but a similar kit of parts: the view is framed by tall trees, contains water in the middle distance and a building on the remote skyline. The picture is contrived to have vertical depth as well as distance.

BLENHEIM

Vanbrugh, Wise, Bridgeman, Brown
1709–1764

Make no mistake, this is a major tourist destination attracting coach loads of visitors from all over the world. Go in summer and find that the car park is full, as is the overflow car park. Marvel at the lengths that managers go to in the interests of squeezing thousands of people through the turnstiles. When the organised parking space is exhausted, cars and coaches alike will be directed towards the central part of the landscape garden to occupy the lawns and drive which sweep down to Vanbrugh's bridge and the Queen Pool. I have seen coaches double parked on the bridge. Most visitors come for the palace experience rather than the landscape design, and the management know that, charging heavily for entry to the former but only modestly for the latter. It is true that access to the palace also includes other elements to the south of the palace but they are not the principal landscape attraction. Henry Wise made a massive parterre there supported by bastion walls, but Brown replaced it with a lawn which some critics consider to be visually unsatisfactory as a platform for the palace. More recently the French landscape architect Achille Duchêne created elaborate water terraces as a foreground to Brown's lake and an Italian-style garden appended to the eastern side of the palace. It would be better to time a visit to Blenheim either outwith the main tourist season, or if this is unavoidable then go in the morning before the park is overcrowded with and disfigured by vehicles.

ABOVE: *Avenues have been used in garden and park design for centuries because they are very powerful and successful on a number of different levels. This low and dark canopied avenue at Ednaston Manor, Derbyshire, focuses on a bright, open space at the entrance court of the house. Direction is important, the detail of the house only being appreciated upon arrival.*

LEFT: *Bramham Park is a layout dependent upon avenues for its composition. Trees are supplemented by beech hedging which has the effect of nullifying the rhythm of tree trunks but strengthens the focus which, in this example is on the distant, undesigned landscape. Grand Manner designs more frequently placed urns, summer houses, sculptures or obelisks as the focal point but Bramham uses both techniques, borrowing the open vista concept from Le Nôtre's central axis at Versailles.*

RIGHT: *The diagram illustrates a hypothetical layout incorporating typical features favoured by Lancelot Brown. The house was orientated for the principal view which itself was enhanced by widening a river or creating a lake in the middle distance. Both the dam with its cascade and the walled kitchen garden would have been screened from the house. The main park frequently had grazing cattle, separated from the immediate confines of the house by a ha-ha.*

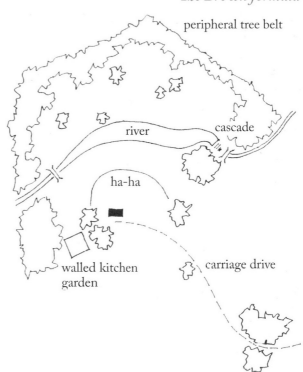

peripheral tree belt

river

cascade

ha-ha

walled kitchen garden

carriage drive

Celebration of level changes

LEFT: *From the nineteenth century, when large estates gave way to smaller gardens and pavings became relatively more important, changes in level were seen as an opportunity to make a design statement. With the cellular gardens of the Arts & Crafts period, level changes were frequently incorporated at the point of space linkage, reinforcing the transition from one garden to another. Circular (convex and concave) steps were common and reflected the use of arches in both walls and hedges.*

The distinctly different architectural styles that Lutyens used at Folly Farm each relate to garden enclosures of appropriate design, here separated visually and physically by an evergreen hedge which is high enough and wide enough to block views from one to the other. This is penetrated by a flight of steps close to the building line, the use of which reveals the climax of the 1912 extension, the cloistered tank court. Whatever the merits of the architecture, which is not without conceit, the handling of the external spaces has been carefully contrived to control what can and cannot be seen from each part of the garden, thus allowing for a simultaneous variety of functions. The banded tubs were not part of the original design concept.

RIGHT: *The 1986 National Garden Festival at Stoke billed itself as The Greatest Event of the Year but much of the construction failed to live up to expectations. Edwardian gardens often included linear water features, or rills, edged with stone laid flush with the adjoining surfaces. In this theme garden, supposed to recreate the spirit of those times and with the use of inappropriate materials badly detailed and failing during the Festival's first week, it became clear that design sensitivity was being sidelined. Three further Festivals had already been committed to by the government but the general standard of design and detailing never approached that at the German Bundesgartenschauen or the Dutch Floriade.*

MIDDLE: *Designers at Glasgow in 1988 contrived to place these bench seats leaning at alarming angles both sideways and forwards.*

BOTTOM: *Sometimes it is necessary to protect corners of lawns or planted areas, although it is always better to design the problem away. If a remedy is required it should certainly be appropriately detailed and if possible enhance the composition.*

Duncombe Terrace

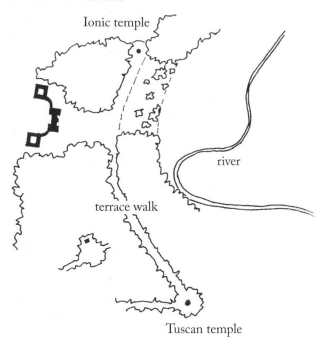

Ionic temple

river

terrace walk

Tuscan temple

The Terrace at Duncombe Park, Yorkshire, incorporates many of the features adopted by the Landscape Gardening movement later in the eighteenth century. Dense woodland frames the house on both the entrance and garden fronts; the Ionic (with its ha-ha) and Tuscan temples at either end of the terrace are treated to similar settings, but one is silhouetted and the other totally enclosed; the curving terrace, which follows the top of the river bluff, allows only one element of the composition to be seen from any location; the planted bluff is variously dense and sparse, controlling the views across the river valley; and there is water in the middle distance.

Exaggerating existing landform

The Howardian Hills are modest in scale, as this prospect from Castle Howard's Temple of the Four Winds demonstrates. Vanbrugh chose to emphasise the landscape's natural qualities by drawing attention to both high and low ground. The mausoleum occupies the highest spot and its dome echoes the gently rolling landform. The bridge, with its reflected arches and relative proximity to the viewpoint, maximises the impression of vertical depth in a landscape that is otherwise largely feature-free. The track linking the two buildings and the deciduous woodland which partially masks the mausoleum both have a role to play in the composition. The billowy form of the wood is entirely appropriate for the scale and character of the wider landscape and the track serves to draw the viewer into the scene and to contemplate on what exactly might be the detail that is hidden by the wood. The overall effect is to double the vertical depth whilst providing some focus in both the middle distance and beyond. Imagine the scene without the buildings and zigzagging track and it might be one that holds a limited interest. This view is taken from the Temple of the Four Winds which is supported on a stone ha-ha.

J C Loudon introduced the Gardenesque approach to landscape design and it proved to be popular as a way of displaying the qualities of individual plants and trees in arboreta and botanic gardens. Characterised by winding paths and free-standing specimens within lawns, a typical example is shown here at Ascott, Buckinghamshire.

Applied to public parks, the Gardenesque included a perimeter belt of shrubbery, for visual containment and screening. Derby Arboretum retains its rather unsubtle if effective landform but has lost much of its fussy detailing like steel hoop edgings to paths. Also, the botanical orders are no longer displayed in its layout.

Here an evergreen hedge set at about eye level allows some details of the garden beyond to be seen but more to be imagined. That the garden is not on show but is clearly cared for and relates to the gabled frontage of the house is at least as interesting as the composition of trees, lawns and borders that might be revealed.

An open gate, this at Little Thakeham, might test the strongest of wills whilst the lowering of a garden wall, where it meets the gable here at Bois des Moutiers, allows partial views of an adjoining and obviously verdant space. Arts & Crafts architects were careful to unite house and garden walls by using the same palette of building materials.

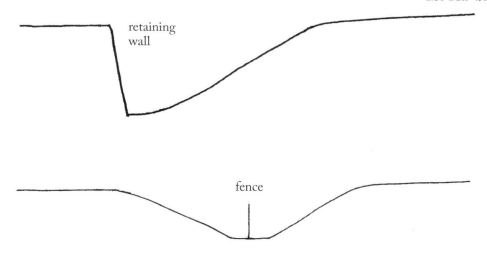

retaining
wall

fence

A ha-ha is a concealed boundary, specifically being concealed from the viewpoint of the property, thus creating the illusion of a garden or parkland much greater in size than the reality of the situation. The device was widely used in the eighteenth century and although some were made by placing a fence at the bottom of a ditch, the most enduring ha-has are effectively stone built retaining walls. They were used to allow livestock in the fields adjoining the property to be contained within the view and to animate it without their gaining access to the immediate surroundings of the house.

The substantial two metres high wall at Bramham Park is one of the best and earliest extant examples.

'The finest view in England' is how the prospect from the Woodstock gate is described in Blenheim's publicity material. This, of course, is a matter of opinion, but it is indeed a very fine view composed of all the classic ingredients which comprised Brown's design philosophy. Coming as a complete contrast to the narrow and usually congested streets of Woodstock, its expansive serenity can take one's breath away. No need for Kodak Photopoint signs here. It is a place of natural pause as first the beauty of the view and later its familiarity sink in (everybody recognises it, don't they?). Regrettably this is as good as it gets. Driving or walking up to the palace, the scene slowly changes bringing a different relationship between the various parts and diminishing the overall effect. There are numerous positions within the park from where bridge, lake and palace can be viewed but none quite as stunning as the one from the Woodstock gate. The components don't vary but their disposition is so cleverly contrived that a near perfect and well-balanced picture is presented to the viewer from that location.

This northern part of the park is visually self-contained whereas Bladon Church, beyond the estate boundary, is brought into the composition on the south side. Surrounding the estate is a backdrop of woodland which helps to focus attention on centre stage. As a playwright, Vanbrugh would have been acutely aware of the importance of this concept but it was Brown who implemented it when his improvements were undertaken half a century later in 1764. Enclosing the site, whether by tree belts and hedges or walls and fences, remains a high priority for many clients today. The requirement to establish an effective enclosure as quickly as possible is not a recent phenomenon, as Henry Wise was instructed by the Duke of Marlborough to create instant maturity with 'full grown trees'. Brown too was working with large specimens but his mass planting of mostly native saplings was a precursor of modern techniques in woodland establishment.

Planned as a triumphal drive, the approach from the Ditchley gate at the north boundary of the estate was never completed, although an avenue was first planted here at the time of the London and Wise design and survived Brown's reworking only to fall victim to Dutch elm disease in the late twentieth century. There is little of landscape interest in this ceremonial avenue. The orientation across the plateau of the Great Park is centred on the Column of Victory, the bridge and the palace courtyard. It is simply an alignment with limited spatial variety and lacks the quality of anticipation or drama that are present at Stowe and Castle Howard.

The Great Park is a good location for a picnic, taken at a leisurely pace, perhaps with some watercolour sketching between courses. In such a manner were two or three hours spent on many occasions by groups of students every spring in the 1980s, fulfilling their role as animators of a Claude landscape painting. Blenheim is not a place for endless exploring: save your energy for Bramham! However, like Hadrian's Villa was for the Grand Tourists in the eighteenth century, it is a 'must do' on the circuit and one which represents extremely good value for money.

From the 1980s a major restoration of the park has been underway, reinstating Brown's tree groups and thereby also the vistas which had been lost over time.

Bowood

Brown 1762–1768, Hamilton 1785

I've chosen Bowood from amongst the numerous parks for which Capability Brown was responsible because, although modified, it represents one of the best preserved in its eighteenth-century form as well as being from Brown's most creative period. At Bowood he was given a blank canvas, a circumstance that doesn't always bring out the best in designers but which Brown would have revelled in, given that he was so frequently called on to improve by modifying the work of others.

The Big House at Bowood was demolished in 1955, leaving the service courts as the residence. Robert Adam worked on both, finishing the first and adapting the second to suit the requirements of the Marquess of Lansdowne. Italian-style terraces were appended to the service court wing in the nineteenth century forming the garden within the park, a complete anathema to Brown and a modification that was not unusual in Victorian Britain. Scarcely a significant property of the English Landscape Garden period escaped the coming of the Italianate terrace during the following hundred years. Bowood also acquired a pinetum in the nineteenth century, since developed further as an arboretum. Sitting behind the house, it doesn't interfere with the visual arrangement of house, lawns and lake, but helps to better define the open areas of the design.

The essence of Brown's work here is the creation of a landscape that appears to be entirely natural, within which cattle grazed in the field between house and lake. A ha-ha, which survives in part, separated the open lawns from the grazing pasture. The lake itself is the consequence of a dammed stream, the gentle contours of the land resulting in a broad body of water backed by a dense wooded area on the far side which continues as a perimeter belt enclosing the park. The simplicity belies the skill of its creation and, as with so many of Brown's schemes, it is the water body that makes all the difference. Wherever he worked, he was the master of this element and recognised that a design without water was always lacking an essential ingredient.

Charles Hamilton was commissioned to create a grotto and a cascade at the outfall of the lake, Brown's earlier designs having been rejected. It's not at all clear why Brown should have been overlooked, given that he had already made a most effective cascade at Blenheim of the same character that Hamilton eventually made here. Both are of the naturalistic rockwork variety and both are very well handled.

Bowood had, for a while, a resident hermit who was paid to live in a hermitage or cave and generally look scruffy. Other estates, notably Painshill, employed 'ornamental hermits' too, but the fad was relatively short lived. Hamilton's advertisement stated:

The hermit must continue on the hermitage seven years, where he shall
be provided with a Bible, optical glasses, a mat for his feet, a hassock
for his pillow, an hourglass for timepiece, water for his beverage, and
food from the house. He must wear a camlet robe, and never, under any
circumstances, must he cut his hair, beard, or nails, stray beyond the limits
of Mr. Hamilton's grounds, or exchange one word with the servant.

For his perseverance, the employee would have received seven hundred pounds
but the first hermit managed only three weeks before giving up the challenge.

OTHER SITES

The name 'Capability' Brown is like gold dust to the owners of landscaped parks
where he worked, or might have worked, or is considered to have influenced the
design. This doesn't necessarily mean that what is left after 250 years has any real
landscape design quality. There will always be academic interest in the work of the
great designers and practitioners, but perhaps it is more important to recognise
those layouts which have survived to display the designer's original intention and
which clearly satisfy both the requirements of the brief and the sensibilities of
the present day. Some parks and gardens which haven't been singled out for their
particular historic or design importance are discussed next.

Harewood, near Leeds in Yorkshire, was a large commission for Brown and later
also for Charles Barry. Brown spent nine years from 1772 creating a setting that
he felt was desirable for the large house that Robert Adam had earlier modified.
The house is of palace proportions and occupies high ground within a vast estate.
Brown, true to his guiding principles, created a lake by damming a river, planted
native trees on a grand scale and left open lawns on the main park front. There's
also a curvilinear drive leading from the arched gateway, which itself is the focus
of the village built for the estate workers. So many owners of historic properties
have had to succumb to the pressures of financial reality since the Second World
War and in order to pay the bills have resorted to adding all kinds of other
attractions. Harewood has not escaped. A Himalayan Garden (there are over a
hundred different varieties of rhododendrons planted at Harewood), an adventure
playground, a café, an ice cream kiosk and a large collection of exotic birds all
occupy their own corner of the grounds. Indeed, during the twentieth century the
character of Brown's design changed significantly, the emphasis switching towards
colourful hybrid shrubs, none of which would have earned a place in the original
design but which Harewood's owners presumably thought would improve their
estate. Gardens are always dynamic and Barry's massive terrace, built in 1844,
demonstrates clearly the fashion of the time, as do the changes referred to above.
It cannot be reasonably argued that they have not diluted the eighteenth-century
concept, which, had it remained true to Brown's plan, today might be considered
one of his masterpieces.

At Audley End in Suffolk, Brown widened the River Cam on the park front of the house and graded the lawns evenly back to the house which is framed by tall and dark cedars. The realigned river bank follows a very gentle S-shaped curve, the line of beauty, and is undisturbed by any marginal planting. Bridges mark both ends of the realignment. The scale is small but the design demonstrates that the same approach he used on major schemes is just as appropriate for modest ones. Audley End was a remodelling job in which avenues and a processional drive were swept away. The house was significantly reduced in size in the early part of the century but it wasn't until 1762 that Brown and Robert Adam were called in to apply a makeover. Sadly, the east or garden front has been filled with a Victorian grass and flower parterre which demonstrates only that Brown's work was so much more successful. The prospect of the house from the B1383, which cuts through the park, reveals the ha-ha, but it's the graceful, three-arched humpback bridge on Spring Hill that provides the perfect foreground foil for the open lawns and west facade of the house beyond. It's a scene that is immediately arresting, a serene idyll that Brown would surely have been proud of if he could have seen the mature product of his work.

In Cambridge, the same river winds its way through six of the university colleges in an area known as The Backs. Although pre-existing as open space at least from the sixteenth century, it was in the middle to late eighteenth century that the landscape began to take on the general appearance which it has today, with poplars and weeping willows planted on the grassed banks. Brown redesigned the western area known as the Wilderness in 1772 and planted many trees, but his proposals in 1779 to make further improvements would have been technically difficult and involved ownership issues as well as the necessary demolition of several bridges. They were not implemented. The landscape is revered nonetheless, with one of the signature views of the city being of King's College Chapel seen across the Cam with its immaculate lawn in the middle ground. Bridgeman had also prepared plans for this area, predictably involving canalising the river.

Alnwick Castle is a property where the private owners have created a new, allegedly Italian Renaissance-inspired, garden on a heroic scale. Fortunately it is well screened from the Brown landscape which provides the setting for the castle on its north and east sides. Open meadows, tree clumps and a widened river combine to make a restful scene which only the castle punctuates. From the elevation of the castle wall, a wide panorama opens up, within which it is quite impossible to detect where Brown's work ends and the rural landscape begins. A greater contrast cannot be imagined between this and the Duchess of Northumberland's bizarre north-facing cascade, wavy hedging, circular fountains, complex ornamental garden, bamboo maze and much, much more.

Floors Castle, outside Kelso in the Scottish Borders, in most respects represents very well a Brownian approach to design, although it is unclear who was responsible for it. It differs from Brown's usual formula in that the design allows the distant landscape within the vista from the house and park and it does so in a most successful way. The properties at both Alnwick and Floors are on high ground from which it is hard to imagine the wider landscape being excluded from the view. Floors Castle itself is as completed by William Playfair and stands on a bluff above the River Tweed with a south-easterly prospect over the confluence of the Tweed and the Teviot. The fertile arable fields here are dotted with single trees and small tree groups but the line of the bluff is marked by dense woodland either side of the house, framing it admirably. William Adam's original estate layout, of which the avenues, ronds-points and rides have all gone, was predictably in the Grand Manner. The ingredients are simple and because the property has not been subjected to Victorian 'improvements' the present composition has a certain purity as well as maturity.

The landscape of northern Northumbria, viewed to great advantage from the A68, has all the ingredients of an idealised English landscape, as if Brown had perhaps been responsible for rather more than estate landscapes. It is no surprise to find that his earliest experience of garden work was at Kirkharle, his home village in Northumberland, and later at nearby Rothley Lake on the Wallington Hall estate. His signature features, like gentle serpentine forms, tree belts and groups, are all there in the Rothley design attributed to him. The same road, as the border with Scotland is crossed, opens up a distinctly different landscape, more rugged, less domesticated and not as well suited to the Brownian improvement formula.

Scotland nevertheless saw the Romantic Revolution, just as it had welcomed the Grand Mannerists beforehand. The Thomas Whites, father and son, were amongst the most prolific designers, producing new estate landscapes between 1770 and 1819, the son never working further north than Fife. James Robertson was responsible for the reworking of the Bruce and William Adam park at Hopetoun House on the Firth of Forth, orientated as it was for a distant view towards North Berwick Law. The most interesting aspect of the layout is that the house is located close to the shoreline yet separated visually and physically from it by the dense plantation of the original plan, the sea presumably being considered too untamed and wild at this proximity even to generate sublime thoughts. It must have been a difficult decision by a landscape gardener to resist the opportunity of opening up the wide view to the Firth of Forth.

Humphry Repton's legacy probably lies more in his modus operandi than in any of the projects for which he prepared designs. Many of his commissions remained unbuilt, possibly because as a consultant, he didn't offer the full design and build service. He drew up plans and perspective sketches, the latter with fold-out sections illustrating the difference between the existing condition and his proposals. Bound in red leather, these submissions to his clients became known as the Red Books. The principle is still in wide use today, although computer graphics have tended to replace watercolour paintings. Repton selected diverse design elements both from his time and from history, without sticking rigidly to a formula. Having started out as a defender of Brown's legacy, he progressed to embrace the Picturesque and was responsible for the introduction of such features as terraces and flower gardens. His philosophy seems to have been to give the client exactly what was asked for, whether or not he thought it was in the best possible taste. Blaise Castle, north of Bristol, where John Nash also worked, retains some elements of Repton's design, including the carriage drive which takes a route through a rugged gorge before emerging from woodland into a light, open area in front of the house. Blaise Hamlet, which followed work on the main estate, was designed by Nash with Repton's son George. Its thatched cottage architecture for estate workers is the epitome of the Picturesque, in a toy-town gothick kind of way. At Milton Abbas, Brown of all people made a supremely picturesque estate village, again with thatched villas but set back from the gently serpentining road and with generous open-plan lawns. At Woburn Abbey, Bedfordshire, Repton's layout included themed sub-gardens, which was a novel development at the time. The park at Sheringham Hall, in Norfolk, was designed and realised along with Repton's son John, and is considered one of his best and most complete surviving works. Repton was responsible for advocating the Mixed Style of garden design, which, in short, did what it says on the tin. Biddulph Grange in Staffordshire, not by Repton, is an example from the 1840s.

Legacy

If landscape gardening can be considered sufficiently distinct from all other garden design styles, then the claim that it is the only original art form that England has contributed to the world of art and design can be well justified. The essential characteristic of idealised landscapes is that they appear to be what they evidently are not. They are an illusion, a romantic version of a perfect place, dependent for their success upon concealing the hand of man. The skills involved in creating these gardens and landscapes are rooted more in art and culture than in science and mathematics and the results are clearly more subtle than anything that was created during the Renaissance. Landscape architects working today may rarely have the opportunity to design grand parks and gardens but they are frequently called upon to use the same skills as their eighteenth-century predecessors by disguising or mitigating the effects of major developments, their knowledge and understanding of the natural landscape being of considerably more help to them than any studies in mathematics or engineering.

Man's understanding of the landscape and his manipulation of it in the laying out of grand estates underwent a fundamental change during the eighteenth century. The process, as has been shown, was incremental and the results differ to a considerable degree. Bowood and Painshill are equal representations of the style because both have more in common than what preceded them. A landscape studded with ruins, temples, grottoes, and so on is one that has been clearly fashioned to present arresting and variable views within a naturalistic setting. However, the simpler layouts of Brown and his followers, where the 'improvements' are not at all obvious, perhaps offer the more lasting legacy into the present age, the works of Loudon and most of those working in the Victorian era having interrupted the natural progression of those traits by at least a hundred years. Brown manipulated the landscape into a credible alternative that could easily be mistaken as naturally occurring, albeit in an idealised form.

Repton was the last designer of note in this period. He believed that his job was to adapt tried expedients to particular circumstances rather than always attempting to provide new solutions. The result of this was that he selectively chose the best of Le Nôtre, Brown and The Picturesque and brought flowers back to the garden, partly to satisfy the new breed of botanical collectors. He preferred his landscapes to have multi-stemmed trees, thickets and a more naturalistic appearance than Brown's copse-like planting. For good reasons, he also extended the zone of artificiality from the house as a transition to the park. He saw that it was rational to accept that the garden and park are different entities and should be separated clearly, by which he meant fenced or balustraded. Terraces replaced ha-has and became elevated viewing platforms – bowing to the seventeenth century as a replacement for the parterre gardens close to the house. In Repton's parks, there were open areas and secluded ones. He allowed the use of all the opposites: avenues and clumps; streams and canals; geometry and nature. In this way it is arguable that he was largely responsible for the climate to develop where the policies of Loudon, Paxton and Kemp could be taken seriously and he certainly showed the way to Charles Barry, whose terraces were retro-fitted to many an eighteenth-century mansion. Having dismantled the aesthetic rules for layout design, the way was clear for those who wanted plants to be placed according to where they grew best rather than where they contributed to the overall effect.

Bridgeman, Kent, Brown, Shenstone, Repton and all the other professional and amateur designers worked to their individual design principles, often producing very different results. The innovative English approach to garden and estate landscape design spread throughout Europe even as it was undergoing stylistic variation in Britain. It is no surprise that European designers were selective from the wide palette available to them under the banner of *le jardin anglais*. In France, two main divisions emerged – those like Stourhead and Painshill which were full of rustic and exotic incident, often alluded to as *anglo-chinois*, and those more in the Brownian style. Both were usually less convincing than the progenitors.

4

SCIENCE, DISPLAY AND PATTERN BOOKS

1830s–1900

Principal designers/writers	Good extant examples	Principal components
J. C. Loudon	*Derby Arboretum*	*Carpet bedding*
Sir Charles Barry	*Birkenhead Park*	*Rusticity*
William A. Nesfield	*Sefton Park, Liverpool*	*Eclecticism*
Sir Joseph Paxton	*Waddesdon Manor, Buckinghamshire*	*Rhododendron dells*
Sir Reginald Blomfield	*Cragside, Northumberland*	*Fountains, band stands, etc.*
William Robinson	*Bodnant, Denbighshire*	*Topiary*
		Terraces

In stark contrast to everything which preceded it, by the Victorian period garden design was no longer in the exclusive purview of the rich and the landed gentry. Influential horticulturists, architects and garden designers described the benefits and sensibilities of their own individual approaches, providing scope for discussion, heated debate and sometimes bitter argument. It follows that design took several directions, more or less concurrently, and nearly always as a counter-reaction to the Romantic Revolution of the previous century. Where art had once imitated nature, it was now to become self-conscious, again.

Three significant 'improvements' are implicated in the changes that took place in Victorian Britain. Plant collectors travelling the world in the first half of the century brought back thousands of hardy exotics, all of which needed to be accommodated in appropriate settings where botanists and the public alike could study and admire them. Consequently, botanic gardens and arboreta flourished. Not unconnected to this, improvements to glasshouse heating, involving water being circulated in pipes from a boiler, allowed many more of the half-hardy and tender exotics to be propagated. These plants, raised under glass then planted out in the garden until the first frosts, became known as bedding-out plants and are typically loud in colour. They were ideal for colour massing and pattern making, being relatively small and therefore two dimensional. In 1827 the lawnmower was invented (patented three years later), putting scythe operators out of work overnight. The new lawnmowers brought an accuracy and facility to the care of lawns that was previously unknown.

With the passing of Repton in 1818, the Scottish polymath John Claudius Loudon assumed the mantle of style guru in relation to gardening, horticulture and arboriculture. He also had distinct opinions on related subjects such as agriculture and natural history. Loudon was essentially a writer of books, journals and papers and his leanings were, consistent with the mood of the age, very much towards a scientific approach. It is largely through his writings that the naturalistic designs of the previous hundred years were steadily phased out in favour of a return to what people generally refer to as formalism. Geometry returned to the environs of the big houses, the terraces promoted by Repton being widely introduced throughout the nineteenth century as adjuncts to eighteenth-century compositions. On these rectangular terraces were elaborate parterres, fountains, statuary and often topiary. Loudon's own interest lay mostly in the urban landscape, on which he gave advice relating to the design of cemeteries, parks and botanic gardens. Indeed, what was for long believed to have been the first ever municipally owned public park, in Derby, was designed by him (a review of this follows later in the chapter). Loudon devised an approach to garden design to which he gave the name 'Gardenesque'. It did little to further the art of landscape design but was and still is adopted widely when the display of botanical specimens is required. It was Loudon's dislike of Brown's 'dull' designs, which Repton at first spent considerable effort defending, and the wide exposure of his theories through the medium of his writings that ultimately prefigured the new approaches to design.

Design detail

This was a complicated period and is perhaps best dealt with by considering the various designers and voices of authority, each of whom brought something different to the mix.

Loudon himself defined the Gardenesque style as follows: 'garden design which is best calculated to display the individual beauty of trees, shrubs and plants; the smoothness and greenness of lawns; and the smooth surfaces, curved directions, dryness and firmness of gravel walks; in short, it is calculated for displaying the art of the gardener'. What he really meant was the displaying of the artlessness of the gardener, for surely his art lies more in composition than it does in the separating out of exhibits, as if they were museum pieces.

The Gardenesque was particularly appropriate for the detailed layout of botanic gardens and arboreta, both of which enjoyed great popularity in the nineteenth century. Winding paths and specimen trees in lawns were the chief ingredients of a recipe for display, without the need for such devices as axiality, relationship to 'the house', or picturesque incident. The genius of the place no longer needed to be recognised.

At the domestic scale of the suburban villa the greatest influence on design came from Edward Kemp, who had been Paxton's superintendent of works at Birkenhead Park and was an exponent of the 'unnatural' style. Loosely, this meant a combination of geometric and picturesque, or the regular mixed with the irregular, and where clipped trees and shrubs could stand next to unclipped specimens. In 1850, in his book *How to Lay Out a Small Garden*, Kemp declared that every house must be regarded as a work of art and there would consequently be a want of harmony in associating it with anything composed or resembling the uncultivated parts of nature. Not short on prescription, his list of dos and don'ts is impressive, only surpassed by the almost universal acceptance of them by the public. In the world according to Kemp there would be no sham buildings, no little surprises, no unsuitable ornaments, no complete enclosure or complete exposure, no artificial mounds, no sculptures, no vases, arbours or baskets for plants, no extreme formality, not too many walks and no very short drives. He approved of the house being approached from a curving drive and for it to have a straight terrace which led by straight walks to a parterre which would be in full view of the house and always to one side of it. Beyond this could be curving walks but not serpentine in form, irregular groups of trees or shrubs at regular intervals, the whole property being surrounded by trees on the broken-belt system. The straight walks were bordered with regularly placed small geometric beds on both sides of the path. Regarding topiary, he favoured only mathematical forms such as globes, pyramids and cones, because all representations of figures or animals were, according to him, in bad taste. Kemp's vision of garden design was widely embraced, spreading through the suburbs at an alarming rate. In the smaller garden, his ideas distilled into:

- a square of lawn with one round bed of geraniums or begonias
- two clipped evergreens, often hollies
- a mixed evergreen shrubbery including spotted laurels
- two urns linked by a balustrade next to the house
- a lean-to glasshouse
- a monkey puzzle tree.

To allow him the benefit of any doubt, Kemp was trying to be helpful in suggesting how to use garden ground to best effect. Unfortunately he wasn't a trained designer so his starting point was always less about what might be appropriate for individual situations than how to accommodate all the various features that he favoured.

The architect Sir Charles Barry was one of the principal proponents of the Italianate revival. He had returned from his Grand Tour, spent mainly in Italy, in the 1820s and started designing both buildings and gardens in the Italian Renaissance style. His first major scheme was Brighton Park, followed by Trentham, Harewood, Dunrobin, Shrubland and Cliveden, amongst many. Shrubland is thought by some to be his finest garden and one which drew direct inspiration from the Villa d'Este outside Rome. At Trentham he employed William Nesfield, who from about 1852 made garden design his main activity. Their design relationship lasted fifteen to twenty years and perpetuated the continental 'formalism' of axial terraces, balustrades, parterres and urns. This was much more of a fundamental return to geometric layouts than the preceding period of Repton and Loudon, where terraces and enclosed gardens were still placed within naturalistic parks. It was a natural reaction to the extremes of The Picturesque and the decline of the landscape gardens in perceived taste, despite the fact that they were only now attaining any kind of planting maturity. There was however no fresh design impulse. It drew strength from Loudon's principle that anything which copied nature (as had the gardens of Kent and Brown) could not be art but was deception. Geometric gardens could never be facsimiles of nature, so, by definition, they could be art. Barry himself claimed that 'the definite and artificial lines of a building should not be contrasted, but harmonised with the free and careless grace of natural beauty'. This could only be achieved by a series of architectural gardens, graduated from regular geometry in the immediate neighbourhood of the building, through shrubberies and plantations less and less artificial, merging eventually with the park or wood. This is precisely the pattern at Trentham, in Staffordshire, where a series of shallow terraces, walled with elaborate balustrades, numerous urns, temple-like porticos, fountains and complex bedded-out parterres, are all set in a Capability Brown landscape. Harewood and Holkham are similar but less extensive. Nesfield used red and yellow brick dust and coloured glass in his parterres to make them interesting in winter, although one of his finest and most simplified works, at Castle Howard, employs a very narrow palette of lawns and clipped yews. Where Barry and Nesfield collaborated the scale of their gardens was generally related to the adjoining countryside. Despite their efforts, there are some crucial differences between the Italianate revival and the real thing, all climate related. In Italy, shade is needed for temperature control and relief from glare. Grass and flowers are both largely absent from the designs. In Britain full use is made of any available sunshine, shelter is necessary and both grass and flowers are much used. In relationship to the landscape beyond the gardens, Barry's designs had to be balanced with the civilised British countryside. To the main Italian ingredients of stone, water and evergreens were added broadleaved trees, herbaceous and hardy plants as well as the glasshouse-reared exotics. Barry's parterres, though richly ornamented, were always related to the building and subordinate to it. This discipline and Barry's working knowledge of the original Italian gardens make his work more assured than most of the gardens where Nesfield worked alone. Queen Victoria herself adopted the Italian style at her Osborne estate on the Isle of Wight, thereby giving it a seal of approval which had a knock-on effect with those property owners making extensions to their country mansions.

Another designer who had tremendous influence was Sir Joseph Paxton. An engineer and proficient botanist, he made his name at Chatsworth, an estate that had been great but was in need of considerable upgrading and modernising. The Duke of Devonshire persuaded Paxton to take on the restoration and extensions there instead of emigrating to America, which had been his original intention. Chatsworth started life as a garden in the Grand Manner, the park later receiving the Brown treatment, and Paxton's improvements were of equal significance to those earlier design statements. He built the Great Conservatory, a masterpiece of glasshouse construction inspired by the botanical structure of the *Victoria regia* water lily. This survived until 1920 when it was replaced by a maze. The camellia houses, Emperor fountain and various rockworks in the garden are also by him. He left behind an eclectic garden composed of different, quite unrelated experiences loosely tied together, but Brown's park survived untouched. Paxton is also remembered for the Crystal Palace created for the 1851 Great Exhibition. His influence extended into public park design, his pioneering Birkenhead Park being the major inspiration for Olmsted's Central Park in New York.

Sir Reginald Blomfield, as well as John Sedding, Harold Peto and others, revived the tradition of Renaissance Classicism, both in his architecture and in his garden designs. He was an outstandingly successful architect and an authority on his chosen style of design. He believed that all garden design should be the work of trained architects, regardless of whether they knew anything about plants. His book *The Formal Garden in England*, 1892, apart from berating the ideas of Robinson, revived the interest in 'formality', his own work being of arguably better quality than the earlier Victorian reproductions. Not a garden designer himself, William Morris liked seventeenth-century gardens which had clipped hedging, topiary and pleached trees but not the prevailing fashion for loud, colourful displays. As an arbiter of taste, his approval can be seen as an important factor in the classical revival taking hold. Peto was devoted to the Italian tradition, which he followed with restraint and a respect for both craftsmanship and materials. He also had a sympathy for and understanding of plants in his designs, in which they played a major part. Though strictly speaking rather late to be Victorian, Blomfield's work can be seen at Mellerstain in the Scottish Borders. Here, in 1909, he reworked the existing plan and made terraces leading to a lawn which falls away dramatically to an irregularly shaped lake. The garden axis focuses on the Cheviot Hills a few miles away to the south, specifically upon the Hundy Mundy, an eighteenth-century folly. In his *Memoirs of an Architect*, of 1932, Blomfield says that the intention had been to build a much bigger garden 'in the best manner of Le Nôtre, but we had to abandon this, indeed it would have required the resources of Louis XIV to carry out the whole of my design'.

William Robinson, a practical gardener and writer, made very few designs himself but advised on how to do it and how not to. He used his first books, on the parks of Paris, predictably to condemn parterres. In 1883 the enormously successful *The English Flower Garden* did much to popularise irregular planting in moderately geometric settings. This is actually the basis for much that has followed, from the gardens of Lutyens and Jekyll to the present day. His essential thesis was *The Wild Garden*, 1870. This was a simple concept which, in his own words, 'is to show how we may have more of the varied beauty of hardy flowers than the most ardent admirer of the old style of garden ever dreams of, by naturalizing many beautiful plants of many regions of the earth in our fields, woods and copses, outer parts of pleasure grounds, and in neglected places in almost every kind of garden'. In short, he was looking to all parts of the garden for their potential to accommodate random plantings, thus making them 'more artistic and delightful'. He listed six reasons why such an approach would be worth following, not least that 'we may cease the dreadful practice of tearing up the flower-beds and leaving them like new-dug graves twice a year'. It is clear from this that he and Blomfield had very little in common in their understanding of what was artful. It was left to Gertrude Jekyll to mediate and see the good in both positions.

Public parks (generally)

Public parks are a Victorian invention, their existence being motivated by social improvements. The expanding industrial cities became overcrowded with consequent little opportunity for relaxing or recreation. In the 1840s the average expectation of life in cities was thirty-two and people sought to spend what recreation time they had in many ways but not in parks. This was a time when one in twenty-three houses in London was a brothel and Britain was busy establishing its empire whilst going through a social revolution at home. Livingstone left to explore Africa in 1840, the year that New Zealand was colonised, Canada was declared a Dominion and Derby acquired an arboretum. The rapid urbanisation in Britain made the nineteenth century distinctive in that for the first time there was active promotion of urban open space. The myth of rural Arcadia fuelled the desire to reconstruct the countryside in the town. The impact of this idea has influenced our approach to open space up to and including the present day. It was believed that open space would contribute directly to the health of the population. The pythogenic theory and theory of miasma held that all disease emanated from filth and putrescence or was due to bad air, so it was thought that access to fresh air would reduce the incidence of disease. Open spaces were considered, quite literally, as the lungs of the city and a positive influence on society, police figures tending to confirm this. One report, following the opening of a new park, indicated that drunkenness and disorderly conduct was down by twenty-three per cent, obscene language down by sixty per cent, gambling down by fifty-eight per cent and summary charges of every class down by twenty-six per cent.

Merseyside was one of the poorest areas on a measure of social deprivation. In 1845 Birkenhead Park was designed by Paxton as a place 'where nature could be viewed in her loveliest garb'. Three years later the Public Health Act gave the responsibility for parks to local authorities: 'Each city should have its large parks', and they were invariably named after the monarch of the day. Liverpool's Sefton Park was exactly what the Act had in mind, being 387 acres in extent. Many boroughs built parks more as prestige symbols than as modest public walks, and Loudon's principles were generally employed. The chief characteristics were:

- Enclosing railings supported by an evergreen screen, usually of privet and spotted laurel.

- A dominant enclosed space of irregular parkland with grass and tree clumps at path junctions, contrasted with geometric bedding-out islands for the display of flower colour in summer.

- Winding perimeter paths.

- Grottoes, cast iron fountains, Japanese gardens, play spaces, shelters, seats, bandstands, pools, tea rooms, aviaries, tennis courts, clock golf, putting greens, etc.

The introduction of sporting facilities at the turn of the century generally had adverse effects on the parks because they were essentially alien in character to the concept of rural Arcadia.

The free time available to most people was restricted to evenings, Sundays and holidays, so pressure for the use of the parks was intense. Consequently they had to be wear-resistant, which led to the rather self-defeating policy of 'keep off the grass'. All intoxicating drinks were rigidly excluded and, to ensure that an acceptable standard of behaviour was maintained, park keepers and park police were employed to break up any games that might have started up. There was something of the genie and the bottle about public parks.

Paxton was the early leading designer of parks, although Thomas Mawson was possibly the best-known designer in the later Victorian period. He had a horticultural background and after designing parks moved on to town planning, advocating the 'composite' form of design. In this he identified four different landscape styles: the completely geometric; the formal; the English landscape style (or more accurately the restrained Gardenesque of Paxton and Kemp); and the natural. His best park work, actually undertaken mostly by his son Edward, is generally agreed to have been Stanley Park in Blackpool, a site dating from as late as 1926 and accommodating a large lake, a central Italian parterre garden, the Blackpool Park Golf Club, a sports arena, cricket pitches, tennis courts, bowling greens, and so on. One example of Mawson's earlier work in Canada, at the coincidentally named Stanley Park, Vancouver, is an exercise in rampant geometry which, although accepted for development, was controversial and not realised due to the outbreak of the First World War.

Derby Arboretum

J. C. Loudon
1840

Contrary to popular belief, Derby was not the first town to own a public park. The Arboretum was designed by Loudon and presented to the Corporation by local manufacturer Joseph Strutt. It is still a public park, the main lines remaining unchanged although planting detail has inevitably been modified from the original scientific collection which Loudon helpfully arranged in botanical order, no doubt with the education of the public in mind. Four years earlier, in 1836, Loudon was commissioned to design a public park called The Terrace Garden, for Gravesend in Kent. Work on that started immediately and the park survived for forty years before the land was sold for development. Like Derby Arboretum, it was designed for both public recreation and education. Many features which were to characterise the standard Victorian park are present at Derby. The design is far from being high quality, but it is interesting that such a small and level site could have had such a major influence in setting a pattern for the rest of the century. It is surrounded by park railings, inside which are evergreen shrubs as a continuous boundary strip. Spotted laurels and cherry laurels do most of the work that oaks and elms did in the previous century. The earth modelling is surprisingly crude, bearing in mind the subtleties achieved by Kent, Brown and Repton. Sinuous mounding is a regular 1.8 metres high and provides effective discontinuity between the central spine footpath and the meandering perimeter paths. However, these abrupt mounds betray their artificial origin too frankly and serve to remind that landform as a component in landscape design is more or less permanent and doesn't mellow with age.

The park is well cared for: bench seating is intact, litter is in the bins, trees are healthy and bare patches in lawns are reseeded (except for the ridge lines which have turned into dirt tracks). It is well used too, located as it is in the middle of a residential area of town. The intimidating presence from at least the 1980s of regular police patrol vehicles is sadly the necessary price for ensuring public safety.

In 1985 my tour notes read: 'Sadly, most of our party don't appreciate why we're visiting this park, despite a lecture course and full briefing about its historical and design significance. Perhaps I should put it all down in print someday – but when do they get time to read?'

BIRKENHEAD PARK

Paxton (with Lewis Hornblower, architect)
1843–1847

Although Derby has the earliest extant public park, Birkenhead soon followed by establishing one through its own auspices. Birkenhead, sitting across the Mersey from Liverpool on the Wirral peninsula, was just being built at the time and the park was included as an integral part of the layout. The design is relatively simple: a serpentining road, Park Drive, meanders around the peripheral areas while Ashville Road cuts the park in two, approximately from north to south. Trees are grouped at entrances, along Park Drive and around the lakes as well as defining the open spaces for organised sports and free play. The simplicity of the plan has been important in its continuing success as a public open space, allowing it to absorb the different requirements of its users. It may not be high art but the ingredients and their disposition within the park are a perfect contrast to the tightly packed urban fabric that surrounds it. It's large enough, unlike Derby Arboretum, for the visitor to be visually divorced from the town, emphasising the contrast. Necessarily, changes have been made to the detailed layout over the years but without materially affecting its essential purpose and arguably improving the quality of its facilities.

Unusually for the early parks, organised sports were always allowed for, with cricket and football pitches both incorporated within the design. Also unusually, there was provision for sheep grazing in one area of the park as a revenue-raising activity.

Birkenhead Park famously inspired Frederick Law Olmsted in his design for New York's Central Park, into which he incorporated many of its features and indeed the principle of the layout itself.

SEFTON PARK

Edouard André, Hornblower
1867–1872

This is one of the largest parks in northern England and is located in the middle of metropolitan Liverpool, between Toxteth and Aigburth. Consistent with Birkenhead and many of the earliest public parks, the perimeter areas were zoned for private housing, the sale of the plots helping to fund the construction of the park. André won the design competition for the park and collaborated with local architect Hornblower, who was responsible for many of the villas, gates, and so on.

Despite his Gallic origins, André's layout was entirely within the English landscape tradition modified by Gardenesque aesthetics, by now the accepted norm for all public parks.

There are large open spaces suitable for any type of activity, organised or otherwise. Tree planting is concentrated along the paths and at junctions where grass areas become too narrow be of any utility. A long boating lake serpentines through the southern area and all the routes through the park, except for one, are sinuous, so contrasting with the regularity of the urban fabric. The public who were lucky enough to be living close by and who found their way into this park would have, and still do have, their own country estate. They have to share it of course, but that is part of the deal, indeed it's the whole point. Sefton is on the same scale as some of those country estates and, like them, from anywhere within its boundaries it is quite possible to imagine that the big city is miles away.

The influence of Parisian parks, also by André, who was a pupil of Alphand, is obvious in his use of circles and ellipses in the layout. The Palm House, influenced by Paxton's great designs and occupying a near central position in the park, was added in 1896 by Mackenzie and Moncur and restored at the beginning of the twenty-first century.

Waddesdon Manor

Elie Lainé, Baron Ferdinand de Rothschild
1875–1889

The entire programme at Waddesdon is ostentatious display writ large. The house, designed by French architect Destailleur, is a Loire château, a composite of parts borrowed from Chambord, Maintenon and others. The garden is high Victorian geometric display with axial symmetry, topiary, hedging clipped to millimetre accuracy, fountains, carpet bedding parterres, aviary, and so on. Every device fashionable at the time is here and paid for by Rothschild money (the bequest to the National Trust in 1957 came with an endowment for its continued maintenance). It's a very bold statement of extreme wealth and it occupies a considerable team of gardeners who maintain it in first class condition.

Waddesdon Manor may be the best example anywhere in Britain of the art of the gardener, at least in respect of keeping things immaculate and true to the taste of its age. Historical references never fail to mention the great effort required to create both house and garden on the site which was, according to the National Trust's leaflet, 'a barren cone-shaped hill' when Baron Ferdinand bought the estate. It continues: 'Extensive alterations had to be made to the terrain and the top of the hill levelled to provide a site for the house.' Restless for his garden to appear fully grown, he used semi-mature trees, most species of which survived despite the technique being somewhat experimental. The entire task is reminiscent of the French Grand Manner approach to landscape design and owes very little respect to the natural possibilities of the virgin site. Rothschild was intent on creating his own slice of renaissance glory, albeit on a relatively modest scale and with Gardenesque characteristics blending with the central theme. More than the garden detail, the house itself stands like a fish out of water in rural Buckinghamshire. Notwithstanding this, Waddesdon was the third most popular National Trust property in the year 2009–2010, with more than 348,000 paying visitors, just behind Stourhead. There is still clearly a fascination or admiration (maybe both) for this style of gardening, as there is for the Stourhead slice of history. They sit at opposite ends of the table and would argue with each other if they spoke the same language, but they don't.

CRAGSIDE
Norman Shaw, Lord Armstrong
From 1869

Working in the second half of the century, Richard Norman Shaw was an architect who was in the vanguard of those who recognised the merits of craftsmanship in design and allied it to the resurgence of interest in vernacular architecture. He may not have had much skill as a garden designer, but his buildings were always robust and well suited for their purpose. The site at Cragside is truly extraordinary, the extreme topography suggesting that it might be the most unlikely of locations for a country house, but here he fashioned a romantic confection that clings to the hillside and dominates with utmost conviction. His client was Sir William, later the 1st Baron Armstrong, whose business became associated with Vickers before merging to create Vickers Armstrong.

One of the devices that the Arts and Crafts Movement architects used was fake ageing. New houses were made to appear old, often by the conceit of deliberately using different architectural styles on the same facade, giving the impression that extensions had been made over time. Cragside is one of the earliest in the genre and Shaw was guilty of using gothic, Elizabethan and Tudor references in the modelling of its exterior. Cragside looks like it has a great deal more history than the fifteen years that it took to build and, from all directions, takes on the appearance of a fairy-tale castle, thanks largely to the exaggerated chimney stacks rising from external walls. If the house is Arts and Crafts, the landscape certainly is not.

The combination of a tracery iron footbridge flying across a deep river gorge, dense and dark evergreen tree planting and romantic architecture paint an image that comes straight out of the Picturesque tradition. It must be emphasised that the trees, as well as the underplanting of rhododendrons, make a significant contribution to the scene's drama, for without the vegetation the site was bleak, rugged and completely unlovable. Rhododendrons flourished in Victorian Britain and here they share the upland acres with other acid-tolerant shrubs like *Gaultheria* as well as heathers and heaths. There are lakes here too, covering over seventy-six acres and providing hydro electricity for the house. There is planting design and there are both paths and drives throughout the wooded estate, but there is no garden design as such anywhere near the house – the land is just too steep. At nearby Knocklaw there are glasshouses and an 'Italian Carpet' garden but these don't relate to the house and its setting. What Cragside delivers, in no short measure, is a magic that the visitor never forgets. The composition is like no other and its simplicity is also its strength. Nowhere else does a coniferous woodland coupled with massed rhododendrons look so good, and they do so only now that the planting has reached maturity. As a measure of the spell that this place can cast, my children, both now fully grown, will still happily confirm that there are indeed trolls living under the bridges deep in the woods.

Much credit must go to Armstrong himself for having the vision to position his house where he did. He was clearly a man who was ready for a challenge and, 140 years later, his efforts have proved to be an unqualified success. The house appears to be ancient, is crammed full of scientific and technical wizardry and stands in a Picturesque revival setting, possibly the finest example of The Picturesque anywhere. This is an example of Victorian eclecticism in its purest form.

BODNANT

Lord Aberconway
1905–1914

The western terraces at Bodnant, for which the garden is justifiably renowned, are in the Italian tradition which was revived in the nineteenth century, but their construction between 1905 and the start of the First World War demonstrates how sometimes fashions change very slowly in garden design. I have included it in this section because it belongs here stylistically and, probably better than any other garden, represents the Italian format as adapted to our northern climes. Barry's garden at Shrubland is perhaps more in the pure Italian style.

Planting at Bodnant actually started in the late eighteenth century, then from 1875 many of the conifers were introduced. It was therefore into a maturing land-scape framework that the terraces were created, albeit at a time when Arts and Crafts compositions were rampant elsewhere in the country. Given the topography at Bodnant, a series of terraces stepping down the hillside and the opening up of a distant view to Snowdonia both seem like eminently sensible solutions to the site. The plan shows how the terrace gardens are axial with the entrance path which passes directly between the south front of the house and the lawn which it faces. In this way the house becomes part of the composition but not necessarily the chief focus. The merging of the geometric layout (house, terraces, glasshouses) with the contour and zigzag paths in the southern part of the site is well handled. This natu-ralistic area, called The Dell, leads down to a minor tributary of the River Conway, and is heavily planted with trees and shrubs that offer a great variety of colour and texture and also accommodates a large rock garden. Elsewhere there are other typi-cally Victorian features, such as a laburnum arch and plant displays in regular beds.

Bodnant offers interest to all garden lovers, whether it be the well-balanced structuring of the site, the wealth of horticultural variety or the individual incidents scattered throughout the garden. In many ways it provides a synthesis of Victorian styles, the individual parts fusing much more sympathetically than they do in many other gardens. The dividing line between pleasure garden and botanical collection is blurred because this garden can be viewed as either, an arboretum providing the structural framework for all within.

Tresco Abbey

Augustus Smith and descendants
From 1834

Although this garden was begun in the mid-nineteenth century, it was Major Arthur Dorrien-Smith who, from 1918, travelled the world as a plant collector, significantly expanding the collection at Tresco.

The climate of the Scilly Isles allows the growing of plants that elsewhere in Britain would be considered tender. Nevertheless, at Tresco Abbey it was realised that some protection would be needed to reduce the impact of salt-laden winds. First, walls were built then shelterbelts planted, behind which the semi-tropical species flourished. Storms in 1987 and 1990 destroyed much of the planting as well as the shelterbelts but these have since been successfully re-established. The gardens are extensive and contain specimens from most corners of the globe, making this a botanical collection but one with few open lawns. The general form is of gravel paths winding their way through massed and deep plantings of palms, agaves and proteas. The Dorrien-Smiths are still adding to their collection which is unique in Britain but, like Inverewe (see below), there is no structure or design control organising the gardens.

Other sites

Dunrobin Castle, just outside Golspie, is the largest house in the Scottish Highlands and occupies a fabulous location on high ground (a rock outcrop) with dramatic long-distance views south and east over the Moray Firth. The impressive multi-avenue approach leads directly from the A9 to the architecturally disappointing north-west side of the castle dominated by a large coach park, which pays all the bills. The seat of the Countess of Sutherland, the present castle has two main periods, the second of which is a stylistic mix of high baronial and French Renaissance which would not look out of place either on a cliff top in the middle Loire Valley or behind wrought iron gates in the Haut Médoc. Superficially it resembles both Château Pichon-Longueville-Baron and Palmer, although surprisingly it just predates both of them, being from 1845. Charles Barry was responsible for this extension of the castle, as he was for the parterre gardens which occupy the low ground below the castle, between it and the sea. At first it might seem somewhat perverse to locate geometric parterres in the foreground of such a splendid panorama, but this is not nearly so awkward an arrangement as the grove of mature trees which separates the two garden areas. In both scale and concept, the idea fails to convince, certainly when viewed from the extreme elevation of the castle itself.

The planting detail is thankfully much quieter than it might have been, or perhaps was, and the maintenance level is as good as is found anywhere, so these redeeming features help to accommodate what is otherwise a stylistic clash. Of the planting, the best is reserved for Portugal laurels which have been used in a very sculptural way forming truly enormous blocks of dense evergreen vegetation through which tunnels and arches are cut. More impressive than the view of the parterres from the castle (a conceit which can only ever be of fleeting interest) is the reverse view, Barry's towers with their exaggerated, pointed caps soaring high above the battered retaining walls that stabilise the rockface. The scale in this case has been judged perfectly. Whether inspired by French or Italian precedents (the guidebook gives conflicting advice), the gardens have the value of surprise, only becoming visible after passing through the castle. The topography, the adjoining woodland and the enclosing high walls all assist in creating a microclimate, within which the Dunrobin garden flourishes.

Barry remodelled the house in 1850 and made an Italian-influenced hillside garden at Shrubland Park, Suffolk, complete with balustraded steps, fountain, parterres and a vista-closing arcade. Mature trees either side of the great staircase have a significant effect on the spatial quality of the composition. Robinson added a wild garden elsewhere in the park later in the century.

Drummond Castle near Crieff has a large Italianate parterre garden with a Scottish twist. As a two-dimensional design, to be seen essentially in plan form from the castle and its terrace, it is notable only for its scale. The fussy, unrewarding layout and detail leave the garden lover wanting a lot more. Even the central feature, at the intersection of paths making a saltire, is misjudged and grossly out of scale. A vista, in the manner of French Renaissance gardens, has been carried across the parterre garden onto the hillside opposite and leads the eye to the horizon, the focus generated by dense woodland either side of the grass passageway. Better Italian-style gardens exist elsewhere in Britain, though it must be said that Italy does them considerably better and always with the inclusion of one essential element which is missing from Drummond: water. It is a bit of a muddle.

Queen Victoria and Prince Albert bought the Osborne estate on the Isle of Wight in 1845 and three years later demolished the house, replacing it with an Italianate confection. The pre-existing eighteenth-century landscape was predictably modified by the inclusion of new terraces on two levels with elaborate parterres of seasonal bedding plants, all very symmetrical and contained within gravel paving. A golf course has since been inserted into the grounds and plans are in hand to restore all the remaining features to their condition at around the turn of the twentieth century. With more reason than most estates of the period, Osborne is peppered with hundreds of specimen trees, many having been planted in commemoration of visits by dignitaries. As a rule of thumb, a Victorian garden can generally be recognised by its evergreen exotics bunched together around the house and skylining from a long way off.

Inverewe, on the coast of the north-west Highlands, is celebrated for its very existence in a hostile environment. The site was prepared with shelter plantings before the development of the garden, which has always been a plantsman's collection with little of interest to designers.

Britain has some excellent botanic gardens, arboreta and pineta, dating from the nineteenth century. Westonbirt Arboretum, in Gloucestershire, was started in 1829 so the original plantings are now mature or over-mature but ongoing works by the Forestry Commission, who acquired the site in 1956, have ensured its continued health and relevance as a collection. The first plantings at Kew Gardens were in 1759, but it wasn't until 1840 that the site was adopted as the national botanical garden, followed soon after by the building of large glasshouses. Heavily treed and with paths winding apparently randomly throughout the park, it fails to make a mark on the history of landscape design. Being the pre-eminent botanical collection obviously sits uneasily with its merits as a park. The Royal Botanic Garden in Edinburgh was established on its present site in 1820 after two earlier incarnations in the city. As might be expected, this too is heavily treed, though the footpath network is rather easier to understand, being arranged more or less in concentric circles, with Inverleith House at the hub on high ground. Its use by the public as a passive recreation area is much treasured, the scientific aspect of its existence being of secondary importance to many.

Buckinghamshire is a small county but has more than its fair share of country house estates and mansions. Ascott, near Leighton Buzzard, was bought in 1873 by Baron Mayer de Rothschild, who lived only a few miles away at Mentmore, a palace-style house by Paxton. A few years later, the house was extended by Leopold de Rothschild, his architect George Devey using a relaxed, half-timbered cottage style which was in the starkest contrast possible to the four-square neo-renaissance opulence of Mentmore. Late twentieth-century changes have introduced new features and remodelled much of the Victorian layout which was completed in 1902, but it retains its eclectic character and gardenesque features, albeit with a modern twist. One of Ascott's saving graces is that it's cellular. The fountains with their attendant carpet bedding are at least well hidden behind evergreen hedges and don't interfere with the splendid view south-east towards the Chilterns. On my first visit I recorded the following passage in my notes: 'This place is everything I had hoped it would be – a marvellous example of Victorian eclecticism – instructive, but not my idea of paradise.' This remains my view, the intervening years serving only to reinforce those first impressions. Ascott is a random collection of diverse set pieces that happen to occupy a site with a great distant view. In my perfect world the house and the view would survive but not much else, the relationship of the house with the landscape being considerably enhanced in the process.

Buscot Park in Oxfordshire is an eighteenth-century layout (1782) into which Harold Peto inserted an Italianate water garden from 1904. It's a little late to be strictly Victorian but it clearly belongs to the movement that began in the nineteenth century. On one of the radiating avenues cut through the wooded area to the east of the house, Peto created a linear water garden which is focused on a temple, also by Peto, on the opposite side of the big lake. There is rhythm in the detailing which is restrained and symmetrical, never overpowering. The park is a strange hybrid, having the layout characteristics of the seventeenth century with avenues, ronds-points and focal features, but also of the Landscape Gardening Movement with the naturalistic lakes and classical temple. A recent addition, more typical of the nineteenth century, is an avenue of monkey puzzle trees.

Robinson's home from 1884 until his death in 1935 was Gravetye Manor in Sussex. There he tried to implement his various theories, apparently with mixed results.

Scotney Castle in Kent is a picturesque ruin set on a moated island, all within an estate of over three hundred hectares which is mostly a mix of parkland and woodland. It was arranged for best picturesque effect sometime after 1837, when the new house was built some way uphill, and the old castle now punctuates the view from a lookout point next to that house. In 1905 the house adjoining the old castle was partly demolished in furtherance of even greater picturesque effect. Even the icehouse, built in 1834, is a curiously formed thatched cone.

Legacy

This was an age not of poets and connoisseurs trying to achieve perfection of style and taste, but of botanists, engineers and scientists using every new device available to achieve ever grander layouts. To some, the nineteenth century was the dark age of design, when Picturesque turned to Gardenesque and when the big idea was to have everything in the garden, the materials themselves becoming supremely important at the expense of the composition.

As with many other aspects of life at this time, gardening and garden design was in a muddle. The landscape tradition deteriorated into winding paths, shrubberies and flower beds, while the 'formal' tradition made its way through the Italianate revival, these two styles finally coming together to produce what some recognised as the best of both worlds. However, the synthesis designs generally show a lack of understanding of the principles underlying either of the great traditions. Gardens were showpieces, prized for their precision and exactness. Like the servants and tradesmen of the day, the begonia and geranium knew their place because it was possible to be exact when laying out exotic, fibrous-rooted plants. The cities grew fast and the people wanted to tame nature once more, in line with the order and politeness standards which the Queen herself set. Everything in the garden was lovely and more marvellous than ever before. Making this possible was the supply of cheap and plentiful labour. Work was judged by effort, so the perception was that the more work expended on laying out and maintaining a garden, the better it was. We now view things a little differently and recognise that there was a general lowering of taste that went hand in hand with a growing desire for ostentatious spectacle.

So, what did the Victorians ever do for us? The parks, botanic gardens and arboreta are still there, most of them anyway, and people still use them although not always in the same way. People still expect to see displays of colourful bedding plants and complain if town hall efficiencies mean that they don't get them. Bedding displays have spread to roundabouts and other publicly maintained spaces, being a highly visible gesture of civic pride, and are often the key element when villages compete in Beautiful Britain awards. The glasshouses are still there. The investment in them was huge at the time and today we are more or less obliged to use our assets, glasshouses being necessary for rearing half-hardy exotics.

There is perhaps still a majority that appreciates and wants Victorian-style landscapes. Aberdeen enjoyed a prolonged period when it seemed always to win *Britain in Bloom*, possibly because of its vast investment in daffodils and begonias. Perth is still a regular winner in the small city category. To an outsider, the beauty competitions between towns, cities and villages seem to be based on the theory that the more colour there is, the better the result. Quantity wins, quality usually suffers. The majority of television gardening programmes further this cause of Gardenesque beauty and have done consistently for over fifty years, programme producers failing to recognise the difference between gardening and garden design, the emphasis always being on flower colour and horticultural techniques.

The invention of the lawnmower led to the development of gang mowers which now deal efficiently with acres of open space, sports fields and lawns and allow us to enjoy the benefits of accessible recreation areas without incurring punitive labour costs.

For the best part of a hundred years the skill of the gardener was much more in evidence than the art of the designer. It took the emergence of characters like William Robinson, John Ruskin and William Morris, with his Craft Workers' Guild, to question the prescriptive designs of Loudon, Kemp, *et al.*, as well as the prevailing mass production ethos of the age, leading eventually to designs of more beauty and utility. Robinson's chief legacy is possibly his popularisation of gardening journals. His lifelong rants against the evils of carpet bedding failed to have much effect, as the practice is sadly just as prevalent today as in his time.

5

ARTS AND CRAFTS
1890s–1920s

Principal designers	Good extant examples	Principal components
Sir Edwin Lutyens	*Folly Farm, Berkshire*	*Geometric room-like extensions*
Gertrude Jekyll	*Deanery Garden, Berkshire*	*of the house*
M. H. Baillie Scott	*Marsh Court, Hampshire*	*Walls and evergreen hedges*
Philip Webb	*Plumpton Place, Sussex*	*Rills and 'quiet' water features*
Sir Robert Lorimer	*Hestercombe, Somerset*	*Cottage garden style planting*
	Ednaston Manor, Derbyshire	*Pergolas*
	Standen, Sussex	

The English are nothing if not nostalgic and sentimental. As the Victorian era was drawing to a close and the Art Nouveau Movement was gaining wide acceptance, so in garden design fashion took a turn away from the epic and grand, moving towards the domestic and handcrafted. There were still country houses being built but those of the Edwardian period were generally smaller and the clients were looking for 'Old England': the manor or the cottage built in a traditional way and surrounded by an old-fashioned garden full of rambling roses, hollyhocks and mulleins. Edward Hudson, editor of *Country Life* magazine, had his finger on the national pulse and, through his correspondent Lawrence Weaver, championed the works of up and coming architects such as Lutyens who were building houses with instant historical appeal. Indeed, Hudson commissioned several country houses from Lutyens and featured them in his magazine, a consequence of which publicity was that his client base grew quite rapidly.

The leading protagonists, Lutyens and Jekyll, lived near to each other in the Godalming area, hence the epithet 'the Surrey School', which is somewhat interchangeable with the term Arts and Crafts when applied to garden design.

Following from Blomfield's comments in the opening pages of his *The Formal Garden in England* (see review of *The Last Country Houses* in Further Reading), architects were taking the lead in the design of gardens around the houses they were building. Notwithstanding this, the landscape architect Thomas Mawson was active at this period and helped with the gardens of some notable buildings but his record shows that he hadn't really tuned in to either art or craft, being much more of an old school, T-square and compass man as well as an ardent follower of The Gardenesque. His admiration of Kemp never waned. Baillie Scott consulted him at Blackwell, in Windermere, Cumbria, one of his masterpieces completed in 1901 and where some terraces with herbaceous borders were made. The result is either admirably restrained or lacking invention. Whichever is more accurate, it is very curious that the house itself is not orientated for the prime views, which are westwards towards Lake Windermere, although the drawing room does have windows set for that view. The Lake District was a popular location for new country houses, many built for northern industrialists and for weekends or holidays rather than as main homes. Apart from Baillie Scott, Voysey built two houses in the same area. Broad Leys is acknowledged as being amongst his best works and is located on a bluff above the lake with direct access to it. The plan doesn't call for a complex series of outdoor rooms, as the prospect to the west commands all the attention, quite rightly. It doesn't take great foresight to plan the house so that the main rooms take in the view, the middle ground being occupied by sloping lawns to the lake. In many ways this is an eighteenth-century solution and it works very well.

William Robinson provided much of the inspiration for Jekyll's own work with plants, whereas Lutyens gained his early impetus from studying the buildings of rural Surrey (the family home was in Thursley). Together they collaborated on about sixty gardens, all country houses, over a span of thirty years or so and their work was nearly always more successful than when they worked alone or with others, the results being a compromise between the extremes of severe classical lines and wild, naturalistic planting. Above all, the scale of most of the houses and gardens was relevant to the early twentieth century, and was accessible to everyone through the pages of magazines and books.

During the birth of the Arts and Crafts Movement in the 1860s there was much cross-fertilising of ideas amongst the cognoscenti as well as grave discussion over such issues as 'What is colour?' and 'What is design?' These intellectual problems occupied Jekyll at the time she was studying at the new Kensington Art School, where the trend of education was more geared towards the decorative arts and crafts than that available through the academies. She studied the paintings of J. M. W. Turner in some depth, particularly his approach to colour and, if Morris had believed in art for art's sake, Jekyll had a more practical perspective. It was only in her old age, when short-sight put an end to most of her art and craft indulgences, that she started to think about gardening. For his part, Lutyens studied architecture at the Royal College of Art before being apprenticed to Ernest George's office in London. At the tender age of 21 he set up his own practice and it was whilst he was working on an early commission in Surrey that he had a chance encounter with Jekyll. The considerable effort that some of his contemporaries, like Voysey and Mackintosh, expended on their drawings and artwork was not Lutyens' way. He put more effort into accumulating a practical knowledge of vernacular building techniques, the principles of which he fully embraced.

Munstead Wood, Jekyll's own house, was to prove a successful springboard for Lutyens' career and in consequence, through his commissions, Jekyll acquired many of hers. Munstead Wood is no longer the garden that it once was, but it has been well documented in Jekyll's own books as well as in other publications. Miles Hadfield suggests that Jekyll's own garden is probably of greater significance in the history of gardening than Stowe.

The clients of the day, as observed by James Boutwood writing for the *Journal of the Institute of Landscape Architects*, February 1969, were often from the aristocracy, or industry, or the City. They wanted small (smaller than the great piles of the nineteenth century) country houses and were generally not content to rely on lawns, evergreens and bedding plants, as the Victorians were. Instant and elaborate settings were required for the ritual of their social lives and all the garden makers of the time were happy to oblige. Boutwood says: 'He [Lutyens] was the first architect of his time to achieve a unity between house and garden through the conscious and deliberate use of architectural forms, projecting the enclosures of the house into its surroundings and positioning and grouping buildings and external spaces in sympathy with the natural shape and feel of the site on which he was working.' This is another way of expressing that he recognised the genius loci and left nothing to chance in the planning of each commission.

If an architect cared for the principles of the Arts and Crafts Movement or was a member of the Art Workers' Guild, it followed that he was probably as obsessive in attention to detail in the garden works as in the house itself. Good detail design is one aspect of a project but is worth nothing without good judgment in the disposition of spaces, whether they be inside or outside. The catalogue of Lutyens/Jekyll gardens includes many where the scale and detail are a perfect fit and it is for this reason that they justifiably hold their place amongst the best of British gardens from any era.

Baillie Scott no doubt spoke for his generation of thoughtful architects in the introduction to his book *Houses and Gardens*, of 1906. He writes:

> For the building and adornment of the house is surely the most important as well as the most human expression of the Art of man. We are apt to consider it in these utilitarian days as a trite formula – a matter of drains, wall-papers, and bay windows – and we are apt to forget the possibilities of beauty which lie in mere building – possibilities which do not necessarily demand great expenditure for their development, but which may be realised in the simplest cottage. Those who dwell amidst the vulgar and impossible artistry of modern villadom may visit now and then some

ancient village, and in the cottages and farm-houses there be conscious of a beauty which makes their own homes appear a trivial and frivolous affair; but such beauty is generally held to be incompatible with modern ideas of comfort and sanitation, and the lack of real comeliness in a modern house is often held to be a necessary concession to practical demands. And so the art of building as practised in modern times is not so much an Art as a disease. In the early stages of the Victorian era it took the form of a pallid leprosy. Nowadays, it has become a scarlet fever of red brick, and has achieved a development of spurious Art expressed in attempts to achieve the picturesque, which in its smirking self-consciousness has made the earlier candid ugliness appear an almost welcome alternative.

He continues: 'It is difficult to know where to turn to escape from this oppressive nightmare of hideous building'. There is a certain amount of 'head in the sand' about this, but on the content and form of gardens he had distinct and well-reasoned views which concentrate on the affordability of such features as bedding displays and lawns. He recognised and advocated the use of 'sombre toned' yews, the depth and intensity of which are the best background for bright flowering plants. He suggests that 'The beauty of the garden will depend to a great extent on its vista effects, and for this purpose its paths must be straight. The ends of these vistas require special attention, and may be treated in various ways – either by a semicircular recess, with a seat, or a summer-house. The garden should not, moreover, be too open and exposed to the sun, but should be full of mystery, surprises, and light and shade.' Some of his statements are less dogmatic and easier to accept than others, particularly when he writes on the sensibility of having lawns and bedding plants in small gardens, where he rightly draws attention to the possibilities of creating good gardens without either. To his great credit, he was acutely aware of the cost of garden maintenance, worrying always about the possible fate of elaborately contrived designs.

It is clear from the designs illustrated by Baillie Scott that architects borrowed detailing freely from their contemporaries. One of his designs for a house and garden in Switzerland has the appearance of a Voysey building with Lutyens fenestration, the garden too being a charming composite of well-tried features.

Design detail

Jekyll's involvement with the planting aspects of the gardens generally elevated them to a higher plane than when architects tried to do everything themselves or weren't interested in getting involved with the garden planning. Jekyll had done a lot of colour research and had developed a deep knowledge of plants and planting techniques, some of which she devised or perfected through her own experimentation. At first she was growing and supplying many of the plants for her schemes, in the manner of a one-stop shop, so her clients could be confident that their money was being spent wisely. Unfortunately, in the later works, many of which she never visited, her choice of plants didn't always suit the climate or microclimatic conditions of the gardens, leading to large-scale failures. It is possible also that some of her clients, although enchanted by her work at other sites, didn't fully appreciate the amount of effort required to maintain what looked like an apparently spontaneous planting scheme. Gardens with a typical mix of hardy perennials, annuals and woody shrubs combining to produce the cottage garden effect required the year-round efforts of several gardeners to keep them in good shape. Labour, particularly following the Great War, was increasingly expensive so many planting schemes fell into a rapid decline or were lost altogether. Indeed, were it not for Jekyll's writings and the survival of many of her plans, her design theories and concepts might be little more than a distant memory.

In matters of colour, Turner's influence on her was possibly greater than any other, although she has also acknowledged the guidance of her Art School tutor, Hercules Brabazon. Her 'impressionist' planting designs (Christopher Tunnard likened her skill with plants to Monet's with paint) invariably involved the blending of colours within a bed or garden enclosure rather than using a monochrome approach, which was a device that gained popularity later in the twentieth century.

Lutyens was obsessed with detail but not at the expense of anything else. The architects' adage 'firmness, commodity and delight' must have been his guiding principle. The building had to be properly built, it had to be suitable for its purpose and it had to raise the spirits. As his gardens were extensions of his buildings, the same applied to them, although there are always exceptions to the rule. Much of his detailing has a historic precedent, often originating from European gardens. The narrow rills are one such feature, the circular convex/concave steps are another. Although we associate the steps device, largely adopted by architects to celebrate a change of level, with the Arts and Crafts period, Dézallier d'Argenville illustrates what he calls 'Stairs at ye End of a Goosefoot' which, although oval in plan, do exactly the same thing. The patte d'oie itself is a device that Lutyens employed regularly, albeit in a scaled-down version.

Jekyll's travels in Spain are thought to have been the source of some of the features that appear not only in many of their gardens but in those by other designers too. Chief of these is the long, narrow rill, contained within stone edging and often incorporating a sculpture or fountain at the mid point of a run. The detail is found at Deanery Garden, Hestercombe and Amport House. Inigo Triggs used it too at Little Boarhunt. Lutyens was fond of creating the illusion of his water features springing from an underground source. He celebrated this by having a round pool at the end of the rill or canal, half covered by an arch of stonework at the top of which a water spout (at Hestercombe they issue from gargoyles) would complete the arc of a hemisphere. Flat arches used at Deanery Garden and Gledstone Hall allude more to a bridge than a natural spring.

Perhaps more than most other architects, Lutyens was apt to mix his paving materials into contrasting patterns. Often this would be white stone edging to red brick in a herringbone layout, but he rarely missed an opportunity to decorate the paths with objets trouvé, like millstones, or to break them into bays that related to windows or doorways. Red tiles were another favourite material, used in paving and walls alike, purely for pattern making.

FOLLY FARM

Lutyens and Jekyll
1906, 1912

This is certainly one of the best planned of their gardens, surprisingly so given that the finished article is the product of two distinct phases of building. A further complication lies in Lutyens' selection of different architectural styles and it's the garden planning and detail that does the job of unifying the two in a way that does not appear fragmented in any way. Few other gardens have quite the same scale development as the spaces connect around the house and it is the variations of scale and mood that retain the interest here. For Jane Brown, author of *Gardens of a Golden Afternoon*, the climax of the sequence is the sunken rose garden which is surrounded by high yew hedges. It is true that it marks the end of a spatial sequence and is a climax in that sense, but the small tank cloister at the interface of the two extensions is a much richer design and the real high point of the composition. Crucially, it is framed by the building and is an inside/outside space with loggias on two sides. The experience of discovering this, the heart of the garden, is enriched by the presence of a yew hedge which is just high enough and just wide enough to visually separate the rest of the garden where there is a more axial treatment than in this homely corner. The large sweep of the second extension roof, simply treated, is all that is visible above the hedge line. Only as the hedge is penetrated, by a short flight of descending steps, does the character of this cloister become evident, with the strong contrast of the arcade set between the planes of the roof and the reflecting water body. It is the hedge, just the right height and width, which makes possible the element of surprise as well as allowing the two characters to co-exist. This skilful articulation of space, richness of detailing and balanced use of materials makes it difficult to draw the line between house and garden and it is therefore fitting that the true climax is a hybrid space.

Like all the gardens they designed, the masterplanning and space division is his, the planting detail hers. In the mid-1980s, when I last visited, Mrs Astor had been making some modifications to the planting detail, replacing plants with grass in the area facing the 1912 extension and placing black and white painted wooden tubs in the tank garden between each of the loggia buttresses. Also, the simple, monoculture planting at the south side of the tank (evident in the historical

Country Life photographs) had been replaced by a mixture of flowering species. Notwithstanding all this, Folly Farm remains a thoroughly delightful setting for taking tea, alfresco on a sunny summer's afternoon, made more special for our party by being served by the hostess and accompanied by home-made cakes. It is a beautiful garden with contrasting spaces and delight at every turn. If I can't have Plumpton Place, there is no other garden anywhere in Britain that I would rather live more than this.

DEANERY GARDEN

Lutyens and Jekyll
1901

In this design, his first for Edward Hudson, both regular and irregular combine to settle the Robinson/Blomfield controversy for good in balancing the principles of both standpoints. The layout is designed to offer a range of experiences from small paved enclosures to the free-form orchard. The principal axis, from the entrance, goes through the house, across the upper terrace and into the orchard. A flight of semicircular steps marks the definition between house-related and orchard-related garden spaces. Each terrace relates to only one face of the house. The fulcrum of the entire composition is the massive chimney stack on the garden front, anchoring the building to the site.

The site design can be rationalised as a fortified style with romantic references. At Deanery Garden the lower terrace is the equivalent to a moat separating the house from the orchard. The upper terrace bridges from the house to link with the orchard. The rill reinforces this idea by springing from a pool that is half under the upper terrace – a device that Lutyens used time and again. Semicircular steps emphasise that the house is elevated above the orchard, where old roses are allowed to ramble. In contrast, pots, tubs and exotics are used on the terraces. It's like a medieval castle with a friendly face.

The key to the success at Deanery Garden is the location of the house on the plot. Lutyens boldly set two wings of a 'U'-shaped house hard up to the wall at the back of the pavement on Thames Street. Three gates in that old brick wall each lead to a garden axis, the central one also punching through the house itself by way of a fabulous vaulted passageway with contrasting brick and stone detailing. The articulation of space and handling of the materials in this cloistered area of the plan is outstandingly good, better than the garden. Overall, Lutyens demonstrates an ability to use the site to maximum advantage, which on most smaller plots is rarely to place the house in the middle.

There have been changes, including a large and quite unsympathetic flat-roofed extension to the house which interrupts one of the three garden axes. It begs the question 'How could this ever have been sanctioned?' On the plus side, a recent project has restored much of the garden to its former glory.

PLUMPTON PLACE

Lutyens and Jekyll
1927–1938

Plumpton Place, an Elizabethan manor house, occupies an island in a lake and, as the final commission from Hudson, was an opportunity for Lutyens to work in the context of large water bodies. The house was fully restored and a music room added, but the project had more to do with the external works and provided the architect with a late flourish in his domestic gardens file.

The moated house is a very English dream and here Lutyens' job was chiefly to sort out the entrance sequence in order to heighten the quality of the pre-existing arrangement. Jekyll had the lake front to play with. Lavish plantings of *Aruncus*, *Astilbe*, gladioli and day lilies occupied the middle ground between the picturesque house and the black swans, quartering their domain and keeping guard like a fleet of mine-sweepers on the upper lake. As in so many of her schemes, the original planting has not survived. If Deanery Garden was alluding to a fortification, this is the real thing. The sequence starts with two symmetrical lodges built on the western boundary of the site, a footway bisecting them on axis with the main door of the house. A small grassed courtyard is made by the flanking wings of the rose-bedecked lodges and serves as a contrast to the drama of Lutyens' timber bridge over the moat. The bridge replaced a narrow causeway and turned a peninsula into an island. The alignment of this axis is not quite a perfect right angle with the house but it doesn't matter because it appears to be so. The geometry of the lodges simply allows insufficient room to achieve a ninety degrees approach, due to the moat's encroachment. The estate is mostly occupied by the lakes which are surrounded by a substantial belt of trees which preserve privacy and provide a backdrop to all the views. It is therefore an enclosed world, a paradise that opens up slowly from the first glimpse at the lodge entry point.

The details in any project matter. They matter a great deal but they are the small print of a contract that the designer has with his project. The big picture, or the programme of the design, orders the buildings and spaces as well as the movement through and between them, addressing the brief in relation to such matters as excitement, discovery, drama, contrast, beauty and above all comfort. Ticking all these boxes should result in high quality and Plumpton Place does this in spades.

After nearly thirty years in the ownership of an American venture capitalist who made sensitive and necessary restorations both to the house and to the garden, the property is once again on the market. One thing is certain: even in such uncertain financial times, there will be no shortage of interest in this house. The most romantic of settings will sell it on its own.

Marsh Court

Lutyens and Jekyll

1901, 1924

The 1920s extension of this house occupies ground that was of little significance in the garden and possibly improves the experience by visually separating the croquet lawn from the intricate and arguably overworked terraces adjoining the south and west sides of the house. For all the complexity of the garden, the house is such a stunning building that it provides the focus of attention from all aspects. It is built of hard, pure white chalk, or clunch, with inlays of flint, tile and red brick, and these materials are also used in the walls and paving that define and decorate the garden spaces. It is clear that nothing, from the massing of the building to the detail of individual squares of flint and the balance of tightly controlled courtyards and open spaces, has been left to chance. It is an experimental yet mature work that simply oozes assurance. Yes, there is something of the 'look at me' about the house, perched on high ground above the River Test, but it belongs firmly to the site and sits well rather than stands proud.

Some of the planting suffered during the time when Marsh Court was a prep school, notably in the sunken pool garden which I found looking distinctly shabby in the mid-1980s. Perhaps surprisingly, the walls and pavings have held up very well indeed. This cannot be said for many of their gardens, where typically the pavings have been given inadequate foundations.

Mention has been made of the complexity of the garden design and it is clear that a great deal more pencil lead was used on the garden plan than on the house, detailed as that is. The sunken pool garden, more than any other space, gives the impression of being over-designed. The containing walls are high, making it feel cramped. An imposing flight of fifteen steps leads directly from water level to tall gateway piers framing the view south, but the scale is too grand for such a small enclosure, access within which is towards the sides. The ratio of width to height is between 4:1 and 4:1.5, where eye level has a value of about 0.5. This translates into a very enclosed space. There's nothing wrong with that in principle but it's very crowded with steps and planting pockets. Normally, comfort levels are lost when the width to height proportions are greater than 4:1 so this space is right on the margins of harmonious balance and is not helped by the complexity of the detail.

Hestercombe

Lutyens and Jekyll

1906

Hestercombe in Somerset was reborn in the 1970s, accurately planted using Jekyll's plans which were found rolled up in the potting shed. Yes, really. The layout dates from 1906 and is Elizabethan in inspiration. It stands apart from and on a lower terrace from the house, which was built and altered in previous centuries. The main area is a square parterre of grass and planting beds edged in stone, walled to east and west and with a pergola closing the southern aspect. Detailing is excellent and a prime example of a garden where the materials obtain a more rustic nature the further they are from the house. Jekyll had great fun planting the walls (there are vertical and horizontal planting plans) and filling gaps in the paving with apparently random daisies and aubretia. Also there are narrow rills, quadrant tanks with water spouts, circular steps, wrought iron gates, and so on – features that are found in many of their gardens. An elm that formed a natural arbour was found to have Dutch elm disease but was rescued by a caring team of arboriculturists. The biggest problem with this garden was that it was built on uncompacted fill. This and the fact that the reservoirs feeding the rills dried up causing the construction to fracture and rills to fail. The restoration has dealt with that problem and the garden delights are now there for all to admire.

It is probably the finest extant example of Jekyll's planting designs, dating from when she was at the height of her powers, and the whole garden is really a celebration of the planting, not the space making. It is all the more remarkable that the planting works so well here, given that Jekyll never once visited the site.

The Victorian house at Hestercombe is ugly. It stands above and looks over the garden and the two are a bad fit. The garden deserves better, much better.

Ednaston Manor

Lutyens
1912–1919 (work interrupted by the war)

Rather like Homewood, Hertfordshire, eleven years earlier, the entrance drive conceals the house until the last minute. A dense, wooded area with underplanted shrubs conceals the house at Homewood until the drive reaches the axis of the entrance front. At Ednaston the approach is through a dark avenue, a side arm of a patte d'oie, which emerges into a light-filled semicircular entrance court, the walls pierced in line with the three avenues.

The house, in four-square classical Queen Anne style and Grade I listed, is the undoubted star and the garden supports it, rather better than, say, at Heathcote or Great Maytham, both of which are imposing buildings that dominate their respective sites. There are terraced gardens on the east and south sides, the former on several levels leading down to a hedged lawn and the latter having many of the usual features that Lutyens was fond of, including twin pavilions. Unusually, there is no room for a signature water feature, its absence selling the composition short in much the same way as at Grey Walls.

The south terrace is paved with red bricks laid to a herringbone pattern and edged with local white Hopton stone. Whilst the pattern making is striking and attractive, it also ties in well with the colours and materials of the house itself. Planting here, when seen by me in the 1980s, was of mostly low-growing herbs, and was both subtle and complementary to the architecture. The terrace is supported by a substantial retaining wall which acts as a ha-ha, keeping the livestock in their paddock. The east terraces are designed to show off effusive planting schemes, each with a particular theme. Climbers and trailing plants help to disguise the high retaining structures and some even make it as far as the house, clothing the walls right up to the eaves. The gardens have, over time, become something of a horticultural collection, with features such as an azalea walk, a rock garden (well hidden behind a yew hedge), a pinetum, a spring border and more. A nursery and plant centre was established there by the late Mr Pickering in the 1980s but has since closed under new ownership. Pickering was a rare individual, keen to share the delights of his house and garden with university students, who in turn were pleased to take advantage of his generosity. Currently the public have no access to the property.

Jekyll was not involved at Ednaston, the original planting having been to the designs of the Player family, who commissioned the project.

STANDEN
Webb
1892–1894

The garden here doesn't bear comparison with the best from Lutyens and Jekyll, neither is it typical of others made by Arts and Crafts designers. The house is the only large one by Webb that survives and the garden as it is today is an overworked original by the landscape gardener G. B. Simpson. His layout was in the Gardenesque style, already out of fashion and thoroughly disliked by all of those who associated with William Morris's views on aesthetics.

However, Webb's handling of the site and his placing of buildings on it resulted in a modest composition that fitted well with the existing farm buildings, forming a seamless union. The owner's wife, Margaret Beale, was largely responsible for the planting and it seems that Webb was content to ensure that the building work was how he wanted it, even if a free hand was never given to him in the layout of the grounds. A wide gravel terrace adjoins the south-facing garden front with its conservatory and is a few steps above the main lawn, divided by a low wall and attendant planting. There is no overt geometric design anywhere and, where Lutyens and others would have celebrated the change of level by making a feature, perhaps related to some focus on the adjacent facade, Webb quietly tucks his steps away at one end of the terrace. The open lawn is the garden focus, bordered by irregular planting of trees and mixed shrubs, and is reminiscent of a park-like solution to enclosure of space. Indeed, loose planting is used as the separating element of each sector of the gardens, including the upper lawn, the bamboo garden, the bowling green and the corner known as Rhododendron Dell. Readers will recognise a family trait with Victorian eclectic gardens, which is further reinforced by a quarry garden and paths that wind through wooded higher ground to the west of the house. In the building, there is also a reference to Norman Shaw's Cragside, with the house being surmounted by a lookout tower.

Standen represents the point where Loudon's way was finally rejected, though the finely interwoven architectural spaces of the mainstream Arts and Crafts gardens were yet to surface. The house here remains rather more important than the garden.

HILL OF TARVIT

Lorimer
1906

Just south of Cupar, in Fife, Lorimer undertook a remodelling task on the mansion house of Hill of Tarvit and laid out a modest garden where the main attraction has always been the prospect to the south-east. Substantial yew hedging is used in conjunction with walling to define the terraces on the south side and a wood surrounds the property to the north, east and west, reinforcing the orientation. Planting has been simplified over time and completely eliminated on the lower terrace. A nine-hole golf course in the middle ground was added in 1924.

The site, which is walled, is rectangular and the house sits in the middle. An even and steep gradient from north to south required the house plot to be levelled but the opportunity to take advantage of the terrain has perhaps not been exploited to the fullest. Steps are treated without the great flourish that might have been expected from some other architects. The design relates well to its site, although the west front of the house, which is where the arrival is made, lacks appeal and there's a suspicion that too much emphasis has been placed upon the view.

LITTLE BOARHUNT

Inigo Triggs
1910

Inigo Triggs, another prominent Arts and Crafts architect, had his gardens featured in Jekyll and Weaver's *Gardens for Small Country Houses* (1911), including this house that he built for himself in Hampshire.

Jekyll admired Triggs as one of 'a small band of people, who by word and deed have shown the right way'. By this she meant that he had studied the history of garden design, published measured drawings and photographs of those which had 'survived the onslaughts of the "landscape" school' and analysed what had made them so worthy. At Little Boarhunt he recreated what might be assumed to be a Tudor-style garden in the courtyard formed by the L shape of his house. The scale is small and the composition very simple.

Included are some familiar elements: rill, garden pavilion, rose parterres, walls, arbours and semicircular steps. Triggs' studies found that these features had been used in Middle Ages Britain as well as on the continent, but his publications post-dated the early work of Lutyens and Jekyll, who discovered history for themselves. What seems peculiar is that one trained in the creative arts can rely so completely on history for the design programme of his own garden. Perhaps he was much more confident when dealing with the form, massing and interplay of spaces in the house.

OTHER SITES

Not everything that Lutyens and Jekyll touched turned to gold. At Lindisfarne Castle, Hudson again employed his favourite designers to refurbish the castle as a holiday home for himself and to make a garden for growing vegetables and cutting flowers. In fact the castle was no more than a derelict shell and there was no garden ground at all. With the building, Lutyens was successful in making it appear to grow from the rock. For the garden, Hudson had his own idea, which was to enlarge a small pond in the field below, turning it into a 'very pretty water garden'. A walled garden was to house a tennis court or croquet lawn. Instead, Hudson got only a small garden, its dimensions arranged for a false perspective, without a water supply and with only a nominal shelter provided by its walls. It stands five hundred metres removed from the castle but facing it across a windswept open field. The concept is good but such a garden can only be a simple mix of plants and paving, undeserving of the praise heaped on it by the National Trust and some others.

There are too many examples of satisfactory gardens by this team to mention here but the main design achievements have been alluded to above.

Probably the most disappointing of all the gardens is Heathcote in Ilkley, Yorkshire. Here, the house and the site planning shout out loud: Private, important, big, expensive. It is pyramid planning with the house (an ugly, wholly unlovable pile) dominating a fussy parterre and oval lawn, both of which would be on sloping ground but for substantial retaining walls. There is no element of surprise or delight. It's simply very, very dull, the entire block plan for the house and garden being mirrored about a central axis. Others, mainly architects, can see great merit in it. Jane Brown includes Heathcote in her 'hallowed two dozen' (*Gardens of a Golden Afternoon*), 'for its Yorkshire ebullience'.

The partnership began properly with the building of Munstead Wood, Jekyll's own house in Godalming, Surrey. In reality the garden there was much more of a plantsman's garden than those which they designed and built together over the succeeding twenty years or so. The house was not incidental to the garden but it had much less of a presence in the planning of garden spaces. To the makers of Sissinghurst, Hidcote, Tintinhull and Crathes, all of whom knew Jekyll, with some having visited Munstead Wood, it made perfect sense that their strongest influence was very likely to be horticultural.

Little Thakeham, in Sussex, has a delightful sequence of spaces which start at the entrance court of a house that sits quite close to its boundary. Projecting walls in

LANDSCAPE AND GARDEN DESIGN

the same stone as the house divide this court from ones on either side, and round-arched gates punctuate them to allow access beyond. There are no great vistas or focal points, merely the contrast of light through dark which impels investigation. Roses and herbaceous perennials flourish in these very sheltered conditions and they decorate the walls and arches as successfully here as anywhere else in the Lutyens collection, but it is interesting that Jekyll had no hand in this project.

Sir Robert Lorimer was a Scottish architect working firmly in the Arts and Crafts style, his designs being rooted in vernacular tradition. His architecture is mostly modest, unflamboyant and respectable but his garden design legacy is disappointingly lightweight.

Earlshall Castle in Fife is probably Lorimer's most celebrated garden design, planted in the 1890s when he was also undertaking restoration works to the castle itself. Unfortunately, it displays very little skill in the field of spatial design. Interestingly, his contemporary, the more inventive Mackintosh, never showed an aptitude for inventive garden design either, as the Hill House in Helensburgh clearly demonstrates. There, the house is a monument surrounded by hedged enclosures, none of which interfere visually with the architecture. The west front of Earlshall extends to become part of the wall which surrounds the garden and separates it from the park, where a ha-ha runs parallel with the drive. In truth, the park and associated mature trees have rather more inherent landscape quality than does anything within the walled garden, where the chief attraction is a quadruple arrangement of topiary yews, each having the plan of a saltire. Like so many topiary gardens, the individual forms have become grotesquely misshapen and difficult to relate to on any level. Topiary as an art form is probably best applied very conservatively, as a focus within a layout, but here it is the central and dominant element of the garden. Walking amongst them, several thoughts came to me over and again: Why fill the garden with random topiary? What purpose are they fulfilling or satisfying? Whatever the answer, the yews at Levens Hall and Earlshall are marginally more interesting to walk amongst than those in the Pillar Garden at Hidcote or the Packwood House collection, both of which are kept in perfect symmetrical order and are therefore boring beyond belief. Packwood, in its defence, does have a mount with a spiral path winding to the summit where a yew tree shelters a garden seat. This represents a very acceptable face of plant sculpture. As I'm writing this it has just been confirmed that RAF Leuchars is to close. The airbase adjoins the Earlshall estate and has been an operational centre recently for Tornados and Typhoons, Lightnings and others before them, so now visitors can look forward to appropriately peaceful visits in the years ahead. Hidcote, a twentieth-century garden, is discussed later. The Packwood topiaries, a nineteenth-century replanting of a seventeenth century original, represent Jesus, the twelve apostles and the multitude, but they fail to deliver anything of design consequence other than the 'sermon on the mount' feature described above.

Lorimer worked mostly in Scotland but spread his wings to the backyard of Lutyens and built two houses in Hascombe, Surrey. High Barn's garden was planned by Jekyll, but it doesn't show. Whilst the form of the house and its materials acknowledge the local vernacular, the garden is another disappointment, particularly when seen in the context of Lutyens' designs. A large, bowling green-like lawn dominates the garden front and is bordered by trees and some herbaceous planting. Virtually next door is the other, Whinfold, where the house is again surrounded by lawns and trees and Jekyll's hand is quite invisible. Spatially, both gardens are very simple and whilst no doubt satisfied their respective clients, the designer's skill is not very evident. Interestingly, Whinfold dates from 1897, the same year that Jekyll was collaborating on her own house with Lutyens and that *Country Life* was first published.

In the 1990s I had the good fortune to be commissioned to prepare plans for the restoration of one of Lorimer's gardens in North Berwick, Marly Knowe. What I had available as documentary information consisted only of the selling agent's brochure from the 1970s. This contained a general description of the garden and some colour photos, one of which was of a completely different house. The garden design is simple, though with rather more structure than those in Hascombe. Double yew hedges arranged axially with the main south door of the house define the herbaceous borders, which are divided by a grass path. The borders appeared to have a scatter-gun approach to their design, with no colour subtlety or height variation. Behind the hedges were, on one side, vegetables and, on the other, roses and cutting flowers. In 1997 I found the hedges to be overgrown by several metres and leaning out from the top. The borders were in a neglected state too and behind the hedges there was nothing. I assumed that, a period of nearly a hundred years having passed, the perennials no longer reflected the original planting design, so a new scheme was devised which had colour swathes arranged in serpentine waves and crossing from side to side over the central grass path. Closest to the house were the loudest primary colours, red and yellow, with some orange mixed in. The progression away to the south went through pink, mauve, blue and pale blue to grey and white. The idea was to ensure harmonisation of adjacent colours and to enhance the effect of aerial perspective, with recessive blues being furthest from the house. The project was only partly implemented when I closed my office and moved away, having left it in the capable hands of a local gardener. Good gardeners are hard to find and I suspect that the owners of Marly Knowe had had some difficulty recruiting in the past. Therein lies one of the problems of this kind of garden: it is only ever as good as the gardener.

C. F. A. Voysey had his own very personal architectural aesthetic, whether working in the Lake District or in deepest Surrey. Most would agree that a building's materials and form should respect its locality but Voysey planted one of his trademark white-harlinged, corner-buttressed houses in the red brick, half-tiled countryside of Haslemere. New Place has a cellular garden, partly geometric or, as the listing text has it, 'formal', but also a wild garden planted up in a manner that must be close to the Robinsonian ideal. It too was worked up with help from Jekyll as well as Sir Algernon Methuen, the publisher and garden enthusiast. Contemporary photographs show a house covered with rambling roses and a garden which was planted skilfully, the upper garden enclosed by a brick wall and containing herbaceous planting and a pond. Lower down, a bowling green and tennis lawn, a wild garden with birch clumps, long grass and beds of flowering shrubs, all very close to the Robinsonian ideal.

In 1897 Voysey built an influential house, Norney Grange, also in Surrey, on a wooded, gently sloping site. The garden front with its twin bays supporting gables and the narrow terrace leading to a lawn both echo Lutyens' Fulbrook House which was being built at the same time. The buttressed corners and harling walls are again the signature features but it might otherwise have been designed by the other architect. Like Lutyens, Voysey was a holistic designer so he can claim all the credit for the successful composition and the detail, but the garden is relatively simple, with some hedging alone defining the three or four grassed enclosures.

William Lethaby was one of the prime movers amongst the Arts and Crafts architects, although his output of country houses was nominal and there is very little evidence of his making much effort with the gardens. His Melsetter House in Orkney (1898) has a small walled garden that seems at first to represent a missed opportunity. However, for what Lethaby felt was his most successful house, the garden was actually splendidly understated. William Morris's daughter May described the place in a letter to a friend as having 'an old garden laid out with daisy bordered beds'. There was also a 'great lawn where seagulls and ravens often sit,' she wrote. The idea of large birds, some black, some white, sitting with their beaks into the wind on a smooth lawn that might have been newly mown in opposite directions, a living draughtsboard of a garden, has a strong appeal and indeed may have been the best solution to the climatic rigours of south Hoy and the Pentland Firth.

Lutyens' collaborator on New Delhi, Herbert Baker, made a grand mansion at Port Lympne, Kent between 1914 and 1920, using the Cape Dutch style. If the house looks out of place, then the garden is certainly out of time, belonging firmly to the Italian revival of fifty years earlier. Baker spent his client's money on numerous terraces and gargantuan flights of steps but only modestly on water features. Russell Page advised on the planting of a double herbaceous border, which was first backed with *Cupressus macrocarpa* and later with Leylandii, the replacement being necessary after a period of thirty-five years of neglect. The garden does not fit happily into any twentieth-century category but it shares certain characteristics with Newby Hall, being compartmentalised and structured around a central spine, the flight of steps. The form seems to be of greater import than the decoration and the view south to Romney Marsh and the English Channel might be of greater merit than anything in the garden itself. The gardens at Port Lympne (pronounced

Lim) have a genuine relationship with the house, so they belong to the site in a credible way, but the chief problem is one of scale. The steps both originate from and lead to nowhere in particular and yet are of such heroic dimensions as might be associated with the Grand Manner. Baker's control of the whole site does at least demonstrate the hand of a designer with a reasonably clear idea. Having started his architectural career in the office of Ernest George and Harold Peto in the 1880s, his earliest influences were the same as those of Lutyens. Craftsmanship was always important for him, although his good judgment has been questioned particularly in some of his later architectural works in India and afterwards.

In the Cotswolds, near the picturesque village of Broadway, M. H. Baillie Scott helped Charles Wade to lay out a garden attached to Wade's Tudor house, Snowshill Manor, in 1920–1923. Together they made a delightful series of enclosures on a steeply sloping site and included ornamentation that had echoes of the sixteenth century. Reflecting the green and white painted garden features at Hampton Court Palace, here Wade Blue was used in an attempt to harmonise with the planting design. Wade was a lifelong admirer of craftsmanship, so it was only fitting that he should have added a garden with terraces and ponds as well as walled and hedged enclosures grouped and linked in the Arts and Crafts manner. The planting may not be of the highest quality, but the space making is arguably more important and has a less rigidly ordered feel than some of Baillie Scott's earlier work, or indeed that of Lutyens, Baker or Triggs. Both house and garden were gifted to the National Trust in 1951.

Lutyens abroad

Although this is a historical review of British landscapes and garden designs, I have determined to break the rules on this one occasion and include a garden that is so significant in the Lutyens panoply that it must not go unrecorded. Le Bois des Moutiers stands within sight of the English Channel at Varengeville-sur-Mer, just west of Dieppe, and is a house remodelling and extension by Lutyens dating from 1898. The house detailing, particularly the fenestration, is unlike any of his other houses except the Ferry Inn on Gareloch in western Scotland, which was built at the same time. Both are a kind of hybrid between Glasgow Art Nouveau and Surrey vernacular – odd perhaps for a Normandy house, but it doesn't look out of context. Lutyens added a set of enclosures to the south facade of the house, separating them physically and visually with immaculately detailed walls and, of course, yew hedging. The walls are modulated to allow glimpse views through to what lies beyond, as do some of the hedges, all of which encourages movement through the sequence. An axis running parallel with the building line cuts across two that are centred on particular features of the house facade, offering options for detours. Typically, every linkage point in the garden is treated with mathematical precision and with well-crafted construction. Steps, paving patterns, pergolas, archways are all employed as means of emphasis but the sense of proportion is of no less importance. It's a strong constructional framework built around an axial plan and Jekyll must have been excited to fill all the spaces with her palette of tried and tested plant groupings. There are trees, mixed borders, rose beds and specimens in terracotta tubs, all on the south side of the house. On the north side she had a large, naturalistic garden, complete with a stream to play with. It was Munstead Wood with bells on.

Le Bois des Moutiers is a splendid visual spectacle, worth the price of the ferry crossing on its own. The house and garden composition is amongst the very best from the period.

Legacy

There has been much written about the Lutyens legacy, mostly by architects and architectural critics about the architecture. For reasons that seem obscure, the Lutyens/Jekyll gardens receive only scant mention in most landscape histories, which concentrate usually on Jekyll's colour schemes with herbaceous material. The real significance of their work together might lie elsewhere. Although Lutyens was not a great innovator, he did have the vision of synthesising all the elements of site design. Jekyll needed the planning and sense of order that Lutyens brought to the table for her schemes to shine. Without it her legacy might have been restricted to her theories and writings. The cult of a Lutyens house with a Jekyll garden is now as strong as ever, even to the extent that their collaboration is advertised when only wished for. Grey Walls, which Lutyens built in 1900, is now a hotel, the brochure of which claims that it stands in a Jekyll garden, despite there being no evidence to support this. Elsewhere, the word 'inspired' (in small print) often accompanies the name Jekyll.

It was perhaps understandable that Lutyens should have been spurned by his contemporaries for having nothing to do with the Modern Movement, but the conservative taste of the British public never embraced concrete, steel and plate glass, preferring still the traditional materials and sense of timelessness offered by the Arts and Crafts period in which he had his roots.

Fortunately, the strength of the Lutyens and Jekyll partnership does not rely solely upon the durability of their gardens. Of course the flowers have gone from some of them, the pavings have crumbled in others and the water systems have occasionally failed. These are all associated with maintenance and, as properties have changed ownership, so the details have often suffered. The plans, Jekyll's in colour, have mostly survived and have been used in the restoration of several gardens. They alone represent a rich resource for the student and provide some compensation for the inaccessibility of some sites. The masterplanning, the detailing, the integration of hard and soft materials and the design programme all repay study. The gardens are rarely symmetrical without having the mystery of interconnecting spaces. Such designs, on different scales, were commonplace in Victorian Britain, denigrating the garden to a display area. They are never overtly artistic but usually very practical.

Each of the surviving schemes has a unique character deriving from the location, the materials and the brief. Each tackles the house/garden relationship in a different way, although there are underlying principles employed in them all. Lutyens studied space, proportion and scale, applying this knowledge very carefully, often with the aim of creating illusion. If the house sits well on the plot and the garden spaces are satisfying entities within the whole, it is a tribute to his application and to his awareness of the importance of rhythm.

In the same way as the house at Bodnant is a part of the composition relating to the garden in a complementary way, so are some of Lutyens' best schemes: Orchards, Deanery Garden, Folly Farm, Little Thakeham and Le Bois des Moutiers. They all exhibit a complex interconnecting of spaces, carefully proportioned and exquisitely detailed to provide a continuity in the various cellular experiences. Invariably these spaces have a direct relationship to a facade of the house, with the sequence beginning in enclosed courtyards and progressing through the Reptonian concept of ever-decreasing formality with distance from the house. The idea was not original, but the scale was different, being much smaller, and the detail is always more fastidious. The best schemes all possess the quality of dynamism. One is led from one area to another, through or around the house, and from enclosure to enclosure until the climax of the composition is reached. It is this rhythm which is lacking in the works of many of Lutyens' contemporaries.

Lutyens and Jekyll had wealthy clients for whom they designed sumptuous houses and gardens. Notwithstanding that most people don't have the resources to employ specialists in this way, the products of their collaboration have real value in terms of planning, spatial composition and detailing which are applicable to almost any landscape design problem today. It's true that the planting schemes were mostly unsustainable then, still more so now, and Jekyll's extreme horticultural jugglings are not particularly relevant to any but the most devoted of gardeners with plenty of time and a private income, but the colour theories are still there to be interpreted with different material as are her many principles relating to the use of planting. Preben Jakobsen, a Danish landscape architect working on the ground-breaking Span Housing projects of the 1950s and 1960s with Eric Lyons and Ivor Cunningham, later established his own consultancy and showed a very skilful ability with herbaceous and mixed borders.

One legacy which has been a long time in gestation is that collaboration between respected practitioners of different professions can be a very good thing. There are still architects who believe that their role is to masterplan everything, with the landscape specialist being called upon only after critical, often irreversible, decisions have been made. Equally, there are landscape architects who don't recognise that architects have anything to bring to the feast except the detailed design of buildings, which must be subjugated to the landscape masterplan. Inter-professional rivalry will always be part of life, as it has been in previous eras, but often the suppression of egos leads to a better result.

The Surrey School built houses and gardens for very wealthy clients and the surviving properties remain in the hands of those who have substantial means. The style has endured. To own such a property is still a dream for hundreds of thousands of art lovers and aesthetes, or those who just nostalgically covet a slice of an idyllic past, but the dream is largely unattainable outwith the ranks of bankers and rock musicians. Of course there is another view of the merits of these Edwardian creations, perhaps held no more strongly than by Jane Jakeman, whose vitriolic article in *The Independent* newspaper (8 April 1995) lambasted the whole movement, blaming pseudo-historical nonsense 'for the traditional cottage-gardens that blot the landscape and deceive tourists'. Jakeman's normal milieu is crime fiction.

6

PLANTSMEN'S GARDENS

Labels are sometimes more trouble than they are worth. Hidcote has been recognised by some observers as being so different (a formal garden with informal planting) that it is worthy of a new classification, hence 'the Hidcote style'. I have grouped Hidcote along with the other examples of the 'outdoor room' type of garden but under the broader banner of 'plantsmen's gardens' because they are all, unquestionably, exactly that. They were made by plant enthusiasts to display their skill and they did so within a structure of outdoor rooms. Great Dixter might well have been included in the Arts and Crafts section, given its origin, but it fits better here because of the way that the plants have become more important than the concept of spaces for different purposes, each related visually and physically to the house.

The common theme running through garden and park design of the early to mid-twentieth century was an attempt to bring colour and variety to all the seasons. Whereas in the Gardenesque style of the nineteenth century each tree or shrub was seen as an individual exhibit, the principle was modified to associate plants with one another, whether in artificial or naturalistic surroundings, in a manner that would demonstrate an artistic skill. Jekyll, more than Robinson, might have been largely responsible for sowing these seeds. The gardens, where the artistry of the plantsman is paramount, take many forms and the groupings that are identified here may be too simplistic for some, too complex for others. Suffice to say that in their own ways all these examples share a universal idea. Unfortunately, the rise in importance of colour and variety often came at the expense of design cohesion. Mention must also be made, briefly, of the craze which swept through Britain at the turn of the twentieth century for rock gardens.

Design detail

Despite the often huge expense involved in building a garden framework and then maintaining the planting in good condition, especially at a time when labour was in short supply and increasingly expensive itself, there were some dedicated individuals, mostly amateurs, who devoted a good part of their lives to the creation of essentially ephemeral works of art. The most fragile are those made in the Jekyllesque style where, in as short a period as two years without regular care and attention, it is possible to lose the garden completely. Those who developed woodland gardens were on safer territory, at least in terms of the timescale, when they might expect that their creations would have a long life.

Following from the tradition of the Edwardian country house gardens, Jekyll's general theories were blended with those of the individuals making their own gardens, as opposed to parks. Borders tended to be straight, set within a geometric pattern which permeated the whole garden, and consisted of mixed planting: perennials, annuals, flowering shrubs and occasional small trees. Jekyll's own patterns based on her colour schemes, which were widely available in print, were copied or adapted. The spring borders and White Garden at Sissinghurst, the Red Borders at Hidcote, even the Golden Garden at Crathes, which appeared as late as 1973 and was based on a plan in *Colour Schemes for the Flower Garden*, all owe their heritage to the same source.

Sometimes with a sub-plot of building a national collection, existing parks were adapted or extended. Popular species were acid-loving shrubs which happened to be floriferous, in other words rhododendrons, azaleas, camellias and magnolias, although heathers also frequently feature in these parks. Trees with interesting bark and/or arresting autumn colour, as well as conifers whose form contrasted with broadleaved woodlands, were included in the mix. Pre-existing pineta and arboreta within the parks and woods were retained for variety at the expense of native trees. Circuitous path networks, often only of grass, were threaded through each area in turn, with bench seating provided at intervals. The aesthetics of these parkland gardens is just as contrived as the outdoor room type. Whereas one has its roots in an imagined idyll of medieval cottage gardens, albeit updated, better ordered and sometimes overtly artistic, the other seems to have derived from a dubious concept that woodland and parks from previous centuries are too boring and need jazzing up.

Colour and texture pervade all these gardens, the best involving a balance between 'busy' and 'quiet' zones, the contrast being important to maintain the interest of those experiencing them.

OUTDOOR ROOMS

Principal designers/practitioners	Good extant examples	Principal features
Lawrence Johnston	*Tintinhull, Somerset*	*Evergreen hedging*
Vita Sackville-West	*Hidcote, Gloucestershire*	*Unrelated themed sub-gardens*
Christopher Lloyd	*Sissinghurst, Kent*	*Plant variety*
	Great Dixter, Sussex	*Year-round colour*
	Crathes Castle, Aberdeenshire	
	East Lambrook Manor, Somerset	

HIDCOTE

Johnston
From 1910

Hidcote Manor itself has a relaxed relationship with the gardens which, apart from having two principal axes at right angles to each other, one of which aligns with the longer wall of the house, seems to have grown rather haphazardly. The original, or 'Old', garden faces the manor house and leading off this is a hedged White Garden with topiary figures which mark the crossing of paving at its centre. The house appears above the hedges, the thatching of the roof as it delineates the upper floor windows echoing the detail of the openings in the yew. It fits well and is a good visual reference. All the individual gardens have some theme or other and Johnston has delighted in incorporating all manner of features into his Gloucestershire hillside estate. Only the Old and White gardens have visual relationships with the house, the others linking one to another without any logical rhythm or escalation of scale. The visitor might quickly ask himself 'Where is this leading to, and why?'

Such questions apply most specifically to the Pillar Garden, where serried ranks of gherkin-shaped yews tower above the border plantings. The central area is grassed, furthering the impression of this being a living version of a ruined and roofless temple. Far from complementing the other spaces in the garden, this area lowers the standard and reinforces the notion that Hidcote is a showpiece for all that it is possible to do with plants, regardless of sensibility or good taste. Equally, although the much-photographed pleached hornbeams of the Stilt Garden are as immaculate as the chestnuts of the Cours la Reine in Paris, in this context they seem to be just another example of effects that can be achieved when the natural growth of trees is interfered with. They are a very recognisable feature of Hidcote but if plant lovers come here to be inspired by the groupings and colour schemes of the various borders, they probably don't rush home and try to incorporate such maintenance-heavy gimmicks as hedges in the sky. The Long Walk is just what it sounds like – a long and unsatisfying walk which is always as empty as the gardens at Courrances, as photographed by Atget at the end of the nineteenth century. Empty, not as there through neglect, because the hedges and lawn are manicured to within an inch of their lives, but surely because they are unloved as a garden experience.

This was the first garden to be accepted by the National Trust, in 1947, and their guidebook recognises the real significance of the place by concentrating heavily on the planting details, or rather by including lists of the plants in each of the sub-gardens. By itself this is perhaps of little value except to those who have a deep knowledge of plants and can recognise what they are looking at. There are some really good plant associations here, Johnston having been inspired by Jekyll and Alfred Parsons, the artist and gardener colleague of Robinson. If the masterplan reflected a better and more logical progression from one space to another and if each sub-garden wasn't such a stand-alone item within the whole, Hidcote could perhaps justify its status as one of the places where the Blomfield/Robinson argument was finally resolved. So many garden histories suggest that this is the case, strangely choosing to disregard the Lutyens/Jekyll era, where house and garden were truly considered holistically.

Hidcote leaves itself open to the criticism that it is the creation of a passionate hobbyist who didn't know where to stop. It has the variety of 'empty' gardens and 'well furnished' ones, necessary to keep the senses alive, but there is perhaps a little too much of the contrived and poor judgment of scale for it to qualify as the work of art that some analysts claim it to be.

TINTINHULL

Price from *c.*1900, Reiss from 1933

Dr S. Price was responsible for laying out the walled gardens centred on the west front of the house, then Capt. and Mrs Reiss developed the area to the north of the house, with the Pool Garden being converted from a tennis court after the Second World War. The earlier works are more satisfying in terms both of scale and of detail, particularly since the early 1990s. At that time one of the most important trees, a majestic old Lebanon cedar, had to be felled owing to disease. Located as it was at the north-cast corner of the garden, this area now lacks any vertical focus, leaving the Cedar Lawn too open and featureless. The transient nature of planting in garden design is perhaps highlighted no more than when such a fine and dominant specimen is lost. Recovery could take a hundred years. The Pool Garden, though separated from the Cedar Lawn by head-high hedging, has suffered by the loss too. The proportions of the space, within which the planting is mostly no higher than the enclosing clements, benefitted greatly from the tree looming high above everything else, balancing the views across the garden and offering important shade. Without it, the entire northern part of Tintinhull's garden is sadly reduced to the status of ordinary. Come back in 2100, when the Baillie Scott edict that gardens should be full of mystery, surprises and light and shade might once more prevail.

The strict geometry of the plan is somewhat out of character with the other 'outdoor rooms' gardens from the beginning of the twentieth century. It is also a good deal smaller than both Hidcote and Sissinghurst, the other principal gardens of the type. In planning terms it has more in common with Arts and Crafts gardens though the detailing is rather simpler and the overall effect is more restrained, less contrived.

Much use is made of planted tubs, placed strategically throughout the garden. Sensitively planted, the tubs themselves are of simple design, mostly terracotta and of a type that garden centres have, in recent years, tended to eschew in favour of highly glazed, richly coloured imports decorated with dragons.

The borders here are mixed, somewhat in the Jekyll manner, though each area has a colour theme, much as at Sissinghurst. Stronger perhaps than the infill planting is the structure. Hedging and punctuation trees are well judged and the clipped evergreens in the first two enclosures bordering the axial path perfectly echo the detailing over the house's west door. This scale and relationship between house and garden is what makes Tintinhull special. It also has one very effective and simple detail, tucked away in a corner under the dark shade of an English yew. A bench seat is placed there for the view across the garden and next to it a square slab of stone, hollowed out to form a bird bath, sits on the ground and brings light into the darkness by reflecting the sky. I enjoyed this little detail so much that I found a stonemason who replicated it for me from a slab of Caithness sandstone. It has travelled with me from garden to garden projecting a flickering light onto the ceiling of dark rooms, thereby bringing the garden inside.

Sissinghurst Castle

Sackville-West and Nicolson
1930

The garden at Sissinghurst is the product of two amateur enthusiasts, each with a different but pertinent skill. Harold Nicolson was broadly responsible for the structural elements, fitting the garden layout in amongst the remains of the old manor and castle buildings. His wife, Vita Sackville-West, filled the spaces with her interpretation of Jekyllesque planting, some of Jekyll's ideas being adopted in whole (primroses as underplanting to the Nuttery); others, like herbaceous borders, she was less keen on. Johnston's Hidcote was much admired by them both, particularly the structure of the enclosures.

Sissinghurst is a cellular plantsman's garden in one of the most romantic of settings that can be imagined. The Elizabethan tower dominates the site, provides a visual focus and reference point, and quite simply looks good. In the disposition of open and enclosed spaces there is a strong discipline that the walls, buildings and hedges provide. Fully half of the garden is given to the orchard which, along with the Tower Lawns, provides a glorious contrast to the exuberant planting of other areas, the most famous of which is the White Garden.

A two-storey range that once formed the front of the old castle, and which was turned into the library by Sackville-West, has centrally located gables framing the covered passageway which is the entrance to the garden. Visual and physical control is strong and is of the type so favoured by Lutyens at Plumpton Place, Berrydown Court, The Salutation and Barton St Mary. A paved walk from the entrance divides an enclosed, grassed courtyard in two and aims directly for the tower, an arched passageway at its centre inviting exploration. The framing of light through dark is a powerful draw that demands attention. Passing through it to the next space, another lawn enclosed by red brick walls and yew hedging is arrived at.

The sequence is full of drama and what little planting there is confined to draping over the walls. The White Garden to the left, the Rose Garden to the right – both offer what most people come for. Straight ahead, the same axis leads directly into the orchard, where the long grass is mown only to define pathways down to the moat which marks the garden boundary. Daffodils in profusion decorate this zone before the trees come into leaf.

The Lime Walk, otherwise known as the Spring Garden, is a narrow space contained by pleached limes and, by way of the Nuttery, leads to the herb garden in the south-east corner of the garden. This particular sequence cuts across the geometry of the Rose Garden and looks very awkward in plan, though not so when experienced at eye level. Designers are advised that if their drawings appear to be well balanced on plan, it can be taken as a promising sign that the design will work on site. There are always exceptions to rules and whilst this plan might not receive top marks under examination conditions, it is a mark of the makers' skill that the shortage of space forcing the hand in this way has been so well concealed. The Lime Walk itself is possibly the weakest piece of design in the whole garden, being over-paved and gaining little or nothing from the pleached limes themselves. The paving does contend with visitors a lot better than grass would, but grass would be a considerably better choice were this a private garden still.

Visitors to Sissinghurst may also climb the tower for a spectacular bird's-eye prospect which takes in the surrounding arable farmland beyond the garden. In spring this is often appropriately very colourful when the oilseed rape is in full bloom.

Taxus baccata, the English yew, once again plays a major part in the successful structuring of a notable garden. It has, over the centuries, proved to be quite the best-performing plant for this purpose, being very dense, providing a dark and even foil for other plants as well as for sculpture, and being responsive to pruning. The enclosures in this garden are essential to its success and, as Sissinghurst is one of the most popular of all gardens in Britain (the car park, which is usually full, is as big as the entire garden), the role of yew hedging must be acknowledged.

This garden is large enough to successfully incorporate different moods, from the big displays of roses or the White Garden's open book of floral celebration to the peaceful orchard with its lawn mown to make pathways through and the adjacent Nuttery close to the tranquil moat. Hidcote is almost exactly the same size.

CRATHES CASTLE

Sir James and Lady Sybil Burnett of Leys
1890s–1973

At Crathes Castle, the castle (1553–1596) stands apart from the geometrically laid-out gardens which are contained within a rectangular walled enclosure of almost four acres. The subdivisions, eight in all, are mostly made with Irish yew hedging planted in 1702 and much sculpted into intricate topiary forms. It normally takes three weeks to prune them back each year.

Jekyll visited in 1895 when there were already well-planted borders, which she reportedly praised for the 'brilliancy of the coloured masses'. This is the same reason that people return to Crathes today, but it wasn't until 1926 when the gardens were really taken in hand by the Burnetts. The most recent addition has been the Golden Garden, made in 1973 from an area previously used as a nursery.

Despite all the effort applied over the years to achieving massed colour displays, the most satisfying element of the garden is unquestionably the yew hedging. It alone makes an impression of quality design when viewed from the castle and, in the context of the croquet lawn, makes a welcome visual relief from the otherwise somewhat overblown profusion of colour in the other 'rooms'. There is an impression that there is too much space available here and that, when compared with the cramped circumstances of East Lambrook Manor, the latter achieves a more pleasing and charming effect.

The National Trust for Scotland have cared for the garden since 1951.

OTHER SITES

East Lambrook Manor in Somerset is a small garden full of variety, both of foliage and of flower colour, with a particular emphasis on variegated species. Plants, which include many old varieties, are packed into beds very tightly and the serendipity of self-seeded annuals is generally allowed to contribute to the overall effect. Margery Fish started this garden in 1937 and since her death in 1969 both the character and detail have been managed in a way that she would have approved of. Paths are very narrow, enclosures high and effective, themed planting dense. Very little, apart from some topiary conifers in one area, appears to have been overtly designed. It is the quintessential cottage garden attached to an appropriately old building.

In 1909 Nathaniel Lloyd bought a fifteenth-century house at Northiam, Sussex, and employed Lutyens to fuse this with a sixteenth-century house of similar character which he had also bought and moved from Kent. The resulting property and garden were, from that time, known as Great Dixter. As might be expected, there is much use of yew hedging, some topiary and many opportunities for herbaceous borders. Nathaniel's son Christopher continued to develop the garden, introducing, amongst other features, a wild flower meadow. Great Dixter's garden has the house at its centre with the various garden rooms rather haphazardly distributed on all sides, with the enclosures being a little irregular and not all arranged to have an obvious relationship with the house. Planting in all its glory is celebrated here.

At Newby Hall in North Yorkshire the Christopher Wren house, which was improved by Robert Adam, stands as the backdrop to some of the longest herbaceous borders in Europe. Yew hedges planted in the early 1920s define the borders, which are 140 metres long, separated by a wide grass path axial with the house. The broader layout of the gardens was apparently inspired by Johnston's work at Hidcote and includes themed mini gardens, each of which is geometric in design. They are planted to provide colour throughout the seasons and will be of interest mostly to horticulturists. The design masterplan, like Hidcote and to a certain extent Sissinghurst Castle, prefigures the layouts of some of the late twentieth-century Garden Festivals, where the various theme gardens are linked to an orientating spine route but have little relationship to each other. Now, in the twenty-first century, the latest generation to inherit the Newby estate have undertaken to stamp their own mark on the garden by digging up parts of the herbaceous borders and changing the plants. It remains a labour-intensive plantsman's garden and one which relies heavily upon hedging to visually separate the various subsections.

PARKS

Principal designers	**Some extant examples**	**Principal features**
No significant figures	*Sheffield Park, Sussex*	*Rhododendrons*
	Savill Garden, Berkshire	*Exotic mix of trees and*
	Knightshayes Court, Devon	*flowering shrubs*
		Lack of focus in design
		Autumn colour
		Variety

SHEFFIELD PARK
A. G. Soames
From 1910

Sheffield Park has a considerable back history which includes some major designers such as Brown and Repton. Arboreta were appearing on many estates during the nineteenth century and in 1885 Sheffield Park saw its own collection started. It was, then, a well-established landscape when Soames bought the estate and started a programme of even heavier planting which continued until 1953. At that point the estate was subdivided and sold off, the National Trust purchasing forty hectares of the park in the following year. This has grown considerably in the intervening years and now the park is over a hundred hectares in extent.

The Trust sells Sheffield Park's attractions on colour, emphasising the spring bulbs, the summer-flowering rhododendrons and azaleas and finally the autumn colours of the many rare trees and shrubs, for which the entry ticket is surcharged. Large lakes with connecting cascades articulate the park and provide important open space within what is essentially a woodland park. Any relationship of today's park to how Brown and Repton left it is long lost, the subtleties of their designs having been subsumed within the twentieth-century lust for rhododendrons and year-round colour. The idealised English landscape has become an exotic composition of rounded colourful evergreen shrubs packed tightly together with columnar conifers which are viewed to best effect across open water. The result of the density and similarity of all the planting is that the park design lacks spatial variety but the Trust must have realised early on that popular appeal doesn't necessarily rest on composition and spatial balance. 'Let them have colour' might have been the mantra at Sheffield Park, and visitor numbers certainly support that, with nearly 200,000 in 2010. It's true that this falls well short of Waddesdon Manor (323,000) and Wisley (803,000) in the same year. It is also true that statistics can prove anything. Sissinghurst, Hidcote, Bodnant, Cragside and Claremont all had fewer visitors.

Autumn colour scenes of Sheffield Park have graced a million calendars and quite probably just as many chocolate boxes, showing that there is a deep-felt longing for the style of landscape that Sheffield Park delivers. It is included here as a good example of the type, but it still has deficiencies, notably a lack of balance spatially.

Savill Garden
E. H. Savill
From 1932

The Savill Garden is a small part of Windsor Great Park, yet still extends to about thirty-five acres. It is a celebration of all types of planting and the authorities used to be a little precious about their creation, issuing rules governing behaviour in the garden. My guidebook from the 1970s advises as follows:

> Visitors are not allowed to pick or touch plants in the garden.

> Visitors are asked not to bring baskets or containers into the garden.

> As a restaurant is provided it is not permitted to bring picnic meals into the garden.

> Children under 16 are only admitted if under the constant care of an adult.

> Visitors are not allowed to bring tripods or easels into the garden without previous permission in writing.

> The garden is not a place for ball games or other such pastimes.

> Visitors are not allowed to use transistor radios or other musical recording machines in the garden. These machines are an irritation to others not modern musically inclined.

It is clear from this list, which has not been reproduced in full, that you may have been allowed to visit the garden but under no circumstance should you enjoy your stay. If you sought plants and trees from across the world grouped together in a Gardenesque-type arrangement, and you were prepared to follow the rules, then you may well have enjoyed your visit. If not, you may have been disappointed. The rules have been relaxed a little now and the plantings have matured. They are dominated by rhododendrons and azaleas, sometimes separately and sometimes together, often with camellias or magnolias. There is also a zone which has herbaceous borders, floribunda and hybrid tea roses in linear beds. A wall, built with bricks salvaged from war-damaged properties of eastern London, faces south and is planted with appropriate climbers and species that favour the shelter and warmth of such a situation.

Managed by the Crown Estate, it is not at all clear why this part of Windsor Great Park was sectioned off and made into a showpiece largely for Himalayan shrubs. It may have been simply that it had not previously been properly incorporated within the park and that its time had finally arrived in the 1930s. I have been unable to track down any design brief, but it seems entirely suitable that Loudon should have been resurrected to guide the planning process. Without a house, or even any normal park features, the design lacks a focus, although the ponds help with orientation to some degree.

KNIGHTSHAYES COURT

Kemp 1870s, Heathcoat-Amory from 1937 (with advice)

A gothic revival house surrounded by an Edward Kemp gardenesque park was inherited by Sir John Heathcoat-Amory in 1937, at which time he and his wife, Lady Joyce, started to redesign and extend the gardens and park. Due to the intervening war years, this process was largely stalled until the 1950s. Lanning Roper, Graham Thomas and Eric Savill all contributed to the reworking and the result is exactly what might be expected.

The Heathcoat-Amorys were very keen to change the character of the estate, disliking as they did the high Victorian layout. Starting with the removal of the bedding-out parterres of the terraces, they moved on to create a Pool Garden where Kemp had left a bowling green, then to make what they called the Garden in the Wood. This entailed the felling of hundreds of Kemp's trees and underplanting the more open result with flowering shrubs and decorative smaller trees. After this the planting was extended into other areas of the park, achieving a unified character of flower and foliage colour throughout the seasons, albeit with a different emphasis in each area. Japanese cherries and maples, dwarf conifers, rhododendrons, willows in variety and many more all have their place.

Nearer the house, where the splendidly clipped yew hedging defines the Pool Garden and the Paved Garden, the planting is of a different character. One *Pyrus salicifolia pendula*, strategically planted by the poolside, provides quite the most dramatic and photogenic situation in the whole garden. The subtleties of the various greens (hedging, lawn, tree foliage) all reflected in the water, and the way that the detailing of this space contrasts with the rampant colour planning of the woodland, only enhances its value. Indeed, for a park where so much effort has been expended on plantsmanship, it is perhaps curious that the iconic image of Knightshayes is of a single pear tree in a peaceful context rather than a riot of massed rhododendrons in full bloom. Lanning Roper was responsible for advising the Heathcoat-Amorys on this particular section of the garden.

OTHER SITES

Exbury Gardens in the New Forest, Hampshire, contains the Rothschild Collection of rhododendrons, azaleas, camellias, and so on, within a woodland setting and a recognisably Gardenesque layout. Spetchley Park, near Worcester, is an estate with a history that starts in the early sixteenth century and has seen improvements to suit the style of each period since. Deer parks are set at one side of the house and to the other side are hedged enclosures containing a profusion of horticultural delight. This interest in variety began in the late nineteenth century and continues to the present day. Many other gardens offering similar experiences are scattered all across the country and are too numerous to list. It would be fair to say that most of them, including Nymans, in Sussex, and Brodick Castle, Isle of Arran, have a very similar emphasis in that they specialise in rhododendrons.

SUSTAINABLE GARDENS

There are very few examples in this category. Hundreds, perhaps thousands, of 'green' or sustainable gardens have been made since the concept became popular but few have become known for the traditional qualities for which people strive to make gardens.

BETH CHATTO'S GARDEN
From 1965

Beth Chatto started to make her two-hectare garden from wasteland adjoining her farm near Elmstead Market in Essex in 1965, the varying soils and drainage on the site providing three distinctly different conditions for planting, thereby determining its arrangement into separate zones. The ground conditions are, in turn, coarse gravel with fine sand, acidic silt, and water-retaining clay. Chatto used these to make a Mediterranean garden, a woodland garden and a water garden, this list having now extended to include several others. In the early 1990s a gravel garden was established where a car park once stood, the planting there having been watered in then left to its own capability in the survival stakes. That she has eschewed irrigation is central to her desire to embrace water conservation. The philosophy underlying the entire development was to use plants in the associations with which they would be found naturally, with the intended consequence being that plants establish and look better in circumstances to which they are naturally adapted. Low maintenance has been a requirement of clients for many decades now and is taken as an essential part of commissions for local authority projects. Chatto had low maintenance in mind for her garden too, notwithstanding the popular clamour for a large variety of specimens that would provide interest throughout the year.

Many people erroneously consider grass to be a low-maintenance surface, whereas mowing a lawn only avoids the gardener having to bend down and get his or her hands dirty in the act of weeding. Chatto demonstrates that it is perfectly possible to garden without grass, although the restfulness of a green lawn can be a very welcome feature when winter comes.

In their own words, Beth and Andrew Chatto 'were able to put into practice the underlying principles of what is now referred to as *Ecological gardening*'. This is a poor title for what is better described as 'sustainable', meaning the maintenance of an ecological balance, and therefore requiring little or no interference by man. At her own garden she used no drawings at all in their layout or development, relying instead on her intimate knowledge of the site.

Chatto's example and her writings have received worldwide acclaim from a public that desires a naturalistic effect without having to pay, in time or money, for ongoing maintenance. What she is promoting is a technique which assumes a design character that imitates nature – a more realistic version of Brown's approach in the eighteenth century. It's a technique that concentrates 100 per cent on plants and their ideal growing substrates, rather in the manner of the association of rhododendrons with acidic woodland soils.

HIGHGROVE

The Prince of Wales, with help
From 1980

The Prince of Wales' garden at Highgrove is something of an anomaly. He declares his environmental philosophy to be that it's better to work with nature than against it, yet the garden is highly organised with avenues, yew hedging and a central axis with close-mown lawns, topiary, pleached trees and a planted pavement. The garden was started in 1980 with some advice from friends. The Prince still calls on others to help with design and detail, each year adding something new to the mix. Tom Turner suggests, in *Gardenvisit.com* (25 October 2009), that this is a post-modern garden that deserves high praise. Whilst Turner himself deserves due respect for his knowledge of landscape and garden design, the epithet 'post-modern' might be more confusing and meaningless than it is helpful. The term was coined by Charles Jencks and applied to architecture, referring largely to the building's envelope. Inside it might be new but it wears antique mix-and-match clothing with a modern, often witty, twist. With a garden it is hard to see how it can be one thing on the outside and another within because the garden experience is only an inside one. Turner admits that Highgrove lacks spatial coherence, which for some would relegate it to a category of 'also-ran', whatever else of significance one might find there. It does certainly reference styles from the past but that is nothing new, as we have seen. The design is a collation of effects from the sixteenth, seventeenth, nineteenth and twentieth centuries, with Victorian and Tudor domination. There is something of the Rothschild garden at Ascott blended with an immature Hidcote about it. Borrowing freely from other styles is nothing more than eclecticism and doesn't need a new descriptive label.

Sustainability at Highgrove means organic gardening, which itself means high maintenance. I expect that he can afford it. If Highgrove is a sustainable garden, it is a different kind of sustainability that Beth Chatto recognises and understands and if the trees could talk back I venture that they might plead to be left alone. Apparently, ten gardeners are needed to maintain Highgrove in a tidily shabby condition.

Whereas in Tudor gardens runner beans were grown on pyramid frames, not for the beans but for their flowers, in this garden it would probably be seen as a win–win situation, for the Prince aims to be self-sufficient in the kitchen garden department and potagers are straight out of the Renaissance handbook of garden design.

It is no surprise that aspects of this garden are slightly weird, given the Prince's views on architecture and town planning. If not for his intervention, London might have had an elegant tower block and new square in the heart of the city designed by one of the leading architects of the twentieth century, Mies van der Rohe. The infamous 'monstrous carbuncle' speech to the Royal Institute of British Architects in 1984, in which he described the proposal as a giant glass stump, led eventually to planning permission being declined.

Legacy

If Beth Chatto's kind of plantsmanship offers the lowest possible maintenance, it finds itself at the far side of the table from the outdoor rooms, dependent as they are on constant attention for their effect. Also sitting at the other end is Highgrove. As a technique the Chatto approach cannot reasonably be applied to very many small gardens whilst still retaining any kind of design element other than in the choice of plants. Once an artificial situation is established (walls, fences, enclosures, paving) the planting element will always need some attention. It may be too successful and require pruning or it may suffer die-back leaving unsightly bare ground. It will inevitably be under fairly close scrutiny. The contrary is true with a sustainable planting design, which relies upon invasion, die-back and survival of the strongest-growing species. The final outcome may be predicted but perhaps not absolutely determined at the outset. In purely design terms this abrogates some responsibility to nature, which is not a problem in itself, but for anyone wanting such a garden, a horticulturist and soil scientist might also be appropriate consultees in addition to a landscape designer.

In the sequence in which the three types of plantsmen's gardens have been set out here, the first is clearly the most contrived and the last might consider itself successful if it were not recognised as being designed at all. The measure of a successful project by professional landscape architects often falls into this latter category. The results are not picturesque or even obvious to a bystander but they achieve success in a quiet way. The planting of a hedge in a strategic position, extending a woodland perhaps, or ground modelling to imitate the natural trends of a site can each contribute to the amelioration of some land or landscape disturbance. Chatto's concept of using appropriate plants to suit pre-existing site conditions is nothing new, except perhaps in the context of a garden and, for it to be effective, such a garden may need to be larger than the average suburban plot.

Hidcote and other 'outdoor room' gardens have, it is widely said, had a major influence throughout the twentieth century. They are certainly very popular, as visitor numbers show, and the British love of flower colour, herbaceous planting and even bedding plants is so ingrained that sustainable garden planting is unlikely to make rapid advances. Heathers and azaleas will continue to be bought and planted in their thousands and will continue to struggle without the regular ministration of peat-based composts to soils that probably contain lime-rich building waste. It seems that no account of the gardens at Hidcote is complete without mention of its inspiration to those with small gardens of their own. Unfortunately none of these accounts actually details what this inspiration is. It is surely not the grand axial planning with tall hedging framing views of the sky, nor the 'empty' space of the open-air theatre, nor even the various pavilions and gazebos. How many small gardens have the luxury of a trickling stream running through them? If small gardens are normally incapable of being subdivided with massive yew and holly hedges, they can at least incorporate lawns, tubs and mixed borders, even the occasional tall tree for punctuation. However, these ingredients are not the sole preserve of one hilltop in the Gloucestershire countryside. Why is it so revered? It isn't the relationship with the house, as we have seen. So, is it the structure or the content – or perhaps both? Spatially, Sissinghurst is much the better, and Tintinhull, as a garden for a home, has a better touch with paving and tubs. Towards the end of the century it became clear that the Hidcote style was not the universal aspiration that many had assumed it to be. Art, not of the plantsman, was beginning to make its presence felt.

7

MODERNISM AND ABSTRACTION

FROM 1930s

When architecture turned to Modernism in the years after the First World War, garden designers were presented with arguably their most challenging task in trying to find an appropriate response. Lutyens and Jekyll had found the formula that perfectly suited the vernacular architecture of the Arts and Crafts Movement, and the various 'outdoor room' gardens were a further development where the planting took centre stage. With Modernist houses, characterised by their stripped aesthetic of building materials used in simple planes, the design of the garden posed difficult questions, but most designers continued to peddle their turn of the century ideas.

Of the landscape designers who found a successful way of marrying gardens to these houses, Christopher Tunnard was widely regarded as the most in tune. He was Canadian born but lived in England between 1929 and 1939, gaining a horticultural diploma and working for Percy Cane, then for the last three years practising on his own before taking up a lecturing position at Harvard. His decision to reject Cane's Arts and Crafts approach to design, labelling it as 'not of our time but of the sentimental past', and to adopt the same principles as the Modern Movement architects had done, came after touring Europe, where he studied avant-garde art and architecture. Cane, incidentally, was a prolific designer and writer who was fond of contrasting columnar and horizontal conifers. He was responsible for the 'garden glade': mown grass enclosed by trees and shrubs with groups of birches set in the grass. Tunnard's articles for *The Architectural Review*, published again later as sections of his book *Gardens in the Modern Landscape* in 1938, became essential reading for a generation of designers. American landscape architects Garrett Eckbo, Dan Kiley and Lawrence Halprin, all masters of their profession, credit the book as a key influence on their careers. Essentially, Tunnard eschewed all the styles that preceded his period, instead electing for a simple statement that he thought was fit for purpose. He can find nothing good or worthwhile in British garden history, possibly because he so strongly believed that his own approach to design was the only sound one. Of the themes or influences on his work, he cites functionalism, modern art and the gardens of Japan from the seventeenth century up to the late nineteenth century (the Yedo period) as the main ones in seeking a new rationalism in garden design. The paradox with Tunnard is that whilst rejecting such styles as the Landscape Gardening school, which could not have developed but for the pivotal influence of art, he too found inspiration in art. At Bentley Wood he also placed the kitchen garden and orchard at some distance from the house, screening them from view. He might have been reading up on his Capability Brown. In a garden design for a country house at Gaulby, Leicestershire, the layout has a peripheral belt of shrubberies and clumps or islands of the same within a lawn, resembling a nineteenth-century public park. With the passage of time this too has simplified and is now down to lawns and trees. Even the paved geometry linking the house and the garden has gone.

Amyas Connell was a Modernist architect who worked in partnership with Basil Ward and Colin Lucas in the 1930s but earlier had an association with Stewart Thomson, with whom he designed High and Over, a country house in Buckinghamshire. Completed in 1931, this house has an open outlook on one side and a concrete terraced rose garden on the other. Like St Ann's Court, by Raymond McGrath, there is a background of woodland which the starkness of the white construction is set against to good effect.

Geoffrey Jellicoe and Peter Shepheard, both founding members of the Institute of Landscape Architects, worked with Modern Movement architects in the 1930s, although the inter-war period and continuing through to the 1950s was a relatively barren one for garden and park design because commissions were rare and the professionals were anyway largely concerned with design in the public realm.

Having worked with both Sylvia Crowe and Brenda Colvin, John Brookes MBE found his niche in the design of small gardens, publishing his influential *Room Outside* in 1969 and following it with numerous other publications as well as designs for small gardens at Chelsea Flower Show. In his early work he used abstract art as his starting point, specifically the works of Mondrian, which he converted from two-dimensional geometry into three and found that a happy balance could be achieved. The difference between Brookes' gardens and those of Jencks (see Chapter 8) is that one uses art to inspire and the other makes art from inspiration. Brookes doesn't need Modernist houses or offices for his gardens to look well and function properly, because the materials he uses and his arrangement of the garden spaces are equally suitable and adaptable to whatever building type or style that he works with. His work has developed stylistically since the 1960s into whatever best suits the house and its location, some of his most impressive designs being for properties abroad. His adaptation of historical detailing is not always handled with the same aplomb. At Denmans, West Sussex, he has made a garden, with a plant centre and café attached. He describes it thus: 'It is a garden full of ideas to take home, and which can then be interpreted within smaller garden spaces'. It bears little relationship to his early work as abstraction has given way to plantsmanship. At the same time as Brookes was building his practice in Britain, Wolf Hünziker, Willi Neukom, Heiri Grünenfelder, Ernst Cramer and others in Switzerland and Anthony du Gard Pasley in Britain were making beautifully detailed small gardens characterised by the sensitive use of hard materials and planting. These were inspired not so much by copying Mondrian and allowing form to dictate function as by a rational use of space and a carefully chosen palette of limited materials, somewhat in the manner of Tunnard. Grünenfelder's garden design in Glarus, realised in 1970 and featured in *Anthos* 1/77, is a model of restraint, using granite paving and squared boulders with ornamental grasses. Gardens like this were the inspiration for a generation of landscape architects who wanted desperately to design gardens but found themselves more often than not at the bottom of the pecking order of planners, engineers and architects and with a mandate to prepare planting plans for horrible triangles of green that were sandwiched between service strips.

In a similar spirit to early John Brookes gardens, but with an altogether more subtle handling, are the gardens of Michael Branch. His own garden at Lephins, Berkshire, from about 1968, is a balanced composition which Peter Youngman described as 'an individual and subtle combination of flowing and rigid patterns that owe no allegiance to past styles or contemporary fashions'.

Bentley Wood and St Ann's Court

Tunnard
Late 1930s

Bentley Wood is in Sussex. Designed for the Modernist architect Serge Chermayeff, who built the house in 1938, the garden here is one of very few that Tunnard made before leaving his fledgling practice for a career in education. Although the building is a strikingly modern range resembling boxy Habitat shelving of the 1960s onwards, the garden plan bears a remarkable resemblance to the south side of Munstead Wood, with grass walks radiating out from the house into an increasingly dense wood. Lawns and specimen birch trees occupied the areas closer to the house. This comparison with Jekyll's own garden is not so far fetched because both accepted the principle of order close to the house being slowly lost to nature in the more remote areas. Not much is left of the garden now. A tennis court and swimming pool have been added and there have been paving and planting changes, some as a consequence of the 1987 hurricane.

St Ann's Court, Chertsey, Surrey, is a bizarre, circular and unashamedly Modernist house where Tunnard lived for a short time before emigrating and he used the building plan as a reference point for aspects of the garden design, which is from 1937. The location is a park that Charles Hamilton designed in the eighteenth century, so Tunnard's work was a remodelling of a mature landscape. An expansive lawn sweeps up to the garden front and disappears as narrow pathways into a surrounding belt of trees. There's a double pond feature to one side which describes part of an arc in plan, but otherwise the garden is very simple and allows the house to dominate.

Unselfconscious and restrained, both gardens prefigure the approach followed by the landscape profession in its formative years, where the hand of man was seen to be at its best when essentially invisible. Largely unstructured spaces like those close to the house in Tunnard's gardens are adaptable for most uses, a strategy that has proved very successful in public park design since the earliest days of the mid-nineteenth century.

Legacy

Ultimately, the British did not fully embrace Modernism, either as an architectural movement or in its gardens. The often brutal concrete of Le Corbusier and the glass and steel of Mies van der Rohe, Walter Gropius and others from the Bauhaus had their influence on British architects in the middle third of the twentieth century, but they struggled to find public favour. At Eaton Hall, Cheshire, a new house was designed for the Duke of Westminster by John Dennys after the earlier one from the 1870s was found to be riddled with dry rot. The uncompromisingly modern result, a flat-roofed, travertine marble clad building, was built between 1971 and 1973 but the aesthetic was disliked in the context of the local landscape and particularly the surviving Victorian chapel. It was therefore given a new face in 1991 and now takes the appearance of a rather dull terrace from middle France. It was indeed a massive clash of styles and the history is interesting particularly in respect of which style emerged as the acceptable face of the English country house. A footnote on the garden: as presently configured it is a grandiose and sterile version of pattern making around strict axial symmetry and is possibly the poorest example anywhere of this style of garden. Space leaks out in all directions. The whole composition is a muddle that relates poorly to the Grand Manner and makes Lutyens' smaller scale Gledstone Hall, in Yorkshire, look like a masterpiece. Tunnard illustrates an equally ridiculous context for a Modernist house, by Peter Behrens, which stands on an Arts and Crafts terrace with rustic brickwork steps.

If landscape and garden designers struggled with Modernism, architects were far from settled in their own discipline. In the catalogue for the Arts Council's *Thirties* exhibition of 1979–1980, Charlotte and Tim Benton point out that there was a lot going on in architecture in that decade: 'Sandwiched between the polar opposites of stripped neo-classicism and constructivism, is a whole gamut of styles from "Wrenaissance", through neo-Georgian, "Stockbroker Tudor" and "by-pass variegated", to "moderne" and Dudock-influenced "modern".' Finding the appropriate response required taking a step back to get a clear view of the big picture, of which Modernism was only a small part.

Tunnard's gardens, few though there are, couldn't be more different from Brookes', yet both designers claim to have gained inspiration from their studies of modern art. Both also responded to the needs of the time. Others, like Jellicoe, were Modernists at heart but found themselves ploughing a different furrow out of necessity. Tunnard's legacy lies less in his deeds than in his writing, even then only his first book. Later ones were focused more on city planning. In a remarkably prescient passage in his summary of British history, he writes:

> The mid-nineteenth century saw the arrival of the professor of landscape gardening. This figure was due to put in an appearance at about this time; Repton had been the perfect prototype, and enough literature had resulted from his period to produce in garden architecture the academic mind. Sad to relate, the professor did not make much stir among the English public … and the academician was sent packing to America where he flourished exceedingly, set himself up in universities, and is no doubt partly the cause of the general lack of originality to be found in the landscape art of that country today.

Tunnard, less than a year later, followed exactly that course himself, first to Harvard then to Yale.

Nothing from 1930s Britain compares to the work of American Modernist architect Richard Meier's parkland setting for his Museum für Kunsthandwerk in Frankfurt, completed in 1985. This book is not about Germany, but there the reader should go if he wants to see how architectural and landscape Modernism can be fused in perfect harmony.

The character of a garden can be altered significantly by the decision to emphasize or relax the signals for movement within it. These two photographs, taken six years apart, indicate that there was a change of heart regarding the narrow paths which had been newly formed in 1983. The invitation to take a closer view of Heyford Bridge at the north end of Rousham's landscape garden is still there in 1989 but is less insisting and perhaps more in keeping with the spirit of the place.

Skilled deception

Apparently random self-seeded trailing plants in dry walls were part of carefully designed, full colour planting plans prepared by Gertrude Jekyll, as here at Hestercombe, Somerset.

The Arts & Crafts gardens drew heavily upon historical themes and were designed chiefly for small scale entertaining of elegant society and the nouveau riche. The visual and physical relationship between house and garden was a constant element but the geometry was always tempered by exuberant planting.

Standen, in Sussex, is an early Arts & Crafts house which dominates its garden. It predates the sometimes fussy over-detailing of terraces, walls, steps and water features of which many of the late nineteenth and early twentieth century architects were guilty.

Fit for purpose

At Chatsworth, Derbyshire, this well balanced and robust flight of steps by Paxton has survived despite the unfortunate loss of the glasshouse with which it was originally associated. It underlines the importance of sound construction being imperative to the durability of garden features.

Public gardens

RIGHT: *The public parks of the mid nineteenth century were designed broadly upon Loudon's Gardenesque principles, which in turn borrowed heavily from the late eighteenth century tradition. Naturalistic forms in both contouring and water features provided the framework for specimen planting and architectural incidents such as bandstands. Belts of trees and shrubs were placed at the periphery, echoing Brown's policy of a hundred years earlier.*

Space linkage

The framing of an entrance or exit indicates quite clearly an invitation to explore. At Stowe, Buckinghamshire, the device is used several times, here linking the open lawns on the south front of the house with Kent's Elysian Fields.

Woodland gardens

Woodland gardens became popular in the late nineteenth century, partly as a reaction to the excesses of elaborate High Victorian display. The constituent parts are simple as is the maintenance regime.

Containers

Container planting suffered through the period of the Romantic Revolution but has been popular before and since. The most successful examples tend to be those with a simple palette of plants and well proportioned tubs. This example is at Levens Hall, Cumbria.

Garden buildings make a vital contribution to the design composition, often being used as a focus. At Castle Howard, Yorkshire, Vanbrugh's Temple of the Four Winds stands at the pivot point linking the terrace walk with the wide vista to Hawksmoor's Mausoleum and beyond. This exquisite building displaying perfect proportions would be sufficient reward without the vista.

Context

At Dunrobin Castle, Sutherland, the Italian style parterre by Charles Barry occupies the low ground and is visible from the castle which itself is perched high above on a rocky outcrop with dramatic views across the Moray Firth. Even with such a dramatic topography, the garden geometry is seen in foreshortened perspective. As an incident, it has more impact and relevance when seen in context like this, being a contrast to apparently free-growing woodland, than it does when walking within the parterre.

The elaborate topiary at Earlshall Castle, Fife, is amusing for children and serves as a playground for hide and seek. The pruning, mowing and edging required to maintain this type of garden in top condition is becoming increasingly unaffordable. Earlshall is one of only a handful that survive in such immaculate condition.

8

THE GARDEN AS ART

FROM LATE 1970s

Principal designers	Good extant examples	Principal features
Charles Jencks	*Scottish National Gallery of Modern Art, Edinburgh*	*Various*
Ian and Sue Hamilton Finlay		
Alain Provost	*Garden of Cosmic Speculation, Dumfries*	
	Little Sparta, Lanarkshire	
	Thames Barrier Park, London	

Late in the twentieth century, a number of talented artists, self-taught landscape designers and trained professionals began to make gardens where it wasn't the plant associations, the sequence of external rooms, the display of sculptures and fountains or even the context of house with garden that was the driving force. Gardens had, since the beginning, been the location for works of art but now they were themselves metamorphosing into the art. All the basic materials are the same but their arrangement and relationship to each other was quite without precedent. Their existence is less about outdoor enjoyment which, in the traditional sense, has sparked the laying out of garden space, but in their extreme form is more to do with challenging the norm in the way of Pop Art or perhaps Shock Art. Some people refer to them as post-modern but, as explained earlier, this is a misleading term and one which has been applied equally to the works of Geoffrey Jellicoe and the Prince of Wales. In the final analysis 'post-modern' might struggle to be recognised as a term describing a cohesive set of garden and landscape designs, but whatever the label, the very purpose of some of these gardens must be considered in a new light.

Charles Jencks has been a leading exponent, having completed several commissioned gardens in Britain and abroad, as well as creating one for himself at his home in Dumfries and Galloway. Jencks uses landform in a way that recalls

Bridgeman at Claremont, but a Bridgeman who has thrown out his mathematical symmetry and turned to heavy drugs.

Graeme Moore has a background in horticulture, art history and literature. Since the mid-1990s he has been making what he calls 'herbe gardens', the name being borrowed from the French for grass. At the International Federation of Landscape Architects' 33rd World Congress in Florence, October 1996, he presented a paper which attracted a fair amount of interest, including mine. At that time he had not branched out into practice on his own account and was using the public stage as a forum for his ideas, as indeed most of the speaking delegates were. The Congress theme was *Paradise on Earth, The Gardens of the XXI Century* and Moore's paper was a late arrival. Sir Roy Strong, writing in 2000, says this of Moore's designs:

> He dispenses with flowerbeds and borders, ornaments, paths, walls and statuary, indeed all those elements which we now look on as essential to garden making. Instead he makes a deeply satisfying garden by articulating the grass into patterns of long and short areas forming circles and squares, axes and cross-axes, sometimes even mowing bold serpentine baroque rhythms. In this he echoes, in late twentieth century terms, voices from the past, for until the plant explosion of the Victorian age, descriptions of some of the most admired gardens never even refer to flowers at all.

Moore himself has noticed how children don't hold back when confronted with his herbe gardens, but 'run into them and run around the patterns laughing and shouting'. Leaving out the usual components is a deliberate ploy which he hopes will demonstrate that they are no longer indispensable. Attractive as it is, the turf maze at Chenies Manor, where the paths are gravel and an obelisk marks the centre, asks fewer questions than Moore's shallow relief grass patterns.

At the same IFLA Congress, the presentation which caught the imagination of the greatest number of delegates and which was a lockout on the day was that of Paula Louise Avery-Winstone of the University of Western England. Her paper was intriguingly entitled *Creating Places that Make People Feel as if they Want to Dance*. As a dancer herself, she turned to dance as 'an alternative catalyst for a design generator … to express aspects of dance in the way that I design landscapes and gardens'. Inspired by the American landscape architect Dan Kiley, she thinks of dance as the movement part of aesthetics. Judging by Moore's observations of how children use his herbe gardens, they may dovetail very well with Avery-Winstone's intentions.

Garden of Cosmic Speculation

Jencks
1988–2003

An architectural writer and theorist as well as a trained architect, Jencks has turned to designing landscapes. At his own home, Portrack House, outside Dumfries, this complex garden of over thirty acres draws on such diverse inspiration as science, mathematics and black holes. There are forty different 'areas' including what have become his signature stepped earthworks but also complex two-dimensional planting pieces in an echo of Tudor and Renaissance gardens.

Jencks has also designed the gardens for several Maggie's Centres. The one at Raigmore Hospital, Inverness occupies a tight space near the hospital entrance and includes some radical grassed earthworks with white gravel paths spiralling to the summits, all of which must be extremely difficult to maintain in perfect order. The garden at the Glasgow Maggie's he describes as a moundette with DNA sculpture and RNA planting and twists in aluminium. The groundsman must dread having to do the mowing.

SCOTTISH NATIONAL GALLERY OF MODERN ART

Jencks
2002

The spacious open lawn that used to occupy the space in front of this art gallery did nothing but separate the building from the road. Jencks' design, with which he was assisted by the architect Terry Farrell, uses three elements: landform, turf and water. He calls it 'Landform Ueda' and says that it is 'based on a strange attractor and the flow of earth and traffic'. The whole piece is sinuous and the landform is emphasised by being stepped. A notice opposite the entrance to the gallery asks politely that the ridge on the main earthwork be avoided owing to excessive wear. The public, even the educated, aesthetically minded public, disregard this request flagrantly. The consequence is that a muddy path interrupts the visual purity of the composition. It may be the garden as art, but it was always intended to be interactive, so these things happen. It is only good until it wears out and then the gallery will have to ask questions about their maintenance budget.

LITTLE SPARTA

Ian and Sue Hamilton Finlay
1978–2006

This garden began its life in 1966 but it wasn't until 1978 when Hamilton Finlay's ideas properly crystallised with the implementation of what he called a 'Five Year Hellenisation Plan'. Little Sparta has been described as a sculpture garden but this is too narrow a view and doesn't do justice to the unique creation of this artist, who dabbled in poetry and philosophy to achieve his artistic expression. The site is a windswept moor above Dunsyre in the southern Pentland Hills, not the most auspicious of locations for a garden perhaps, but it suited perfectly Hamilton Finlay's cause.

Nowhere else can be seen such a concentration of concrete poetry, most of which feeds classical allusion but there is also wit in the wordplay. Every artefact has been located with the composition of the whole garden in mind and the central idea of mood creation relies heavily on the treatment of planting, which is mainly the work of Ian's widow Sue. The final work at Little Sparta was an extension of the garden downhill from the house. Here, in what Finlay called the English Parkland garden, he used uncut ribbons of grass to describe a path within a lawn. Beyond the 'path' are trees and shrubs planted in a way that recalls Loudon's Gardenesque. The investment in maintenance will always be greater in this extension area than in the confines of the original works.

THAMES BARRIER PARK

Alain Provost (Groupe Signes)
1996–2000

Twenty-two acres of riverside dereliction has been reclaimed to make this park on the north side of the Thames Barrier. The design is formed around a bold geometric pattern which bears a strong similarity to Provost's earlier Parc André Citroën in Paris (1992) and has been described variously as post-modern or abstract. A sunken linear trench slices through the site and contains the Rainbow Garden which is characterised by lines of complex hedging with undulating tops, alternating with strips of coloured plants. High-quality and expensive maintenance will be required if this zone is to retain its sharply formed detailing.

The form and geometry of this type of park design would have been highly appropriate as a setting for Modernist architecture, the extremes of which had already been consigned to history long before 1996.

The main concern of the design team was to 'create a rich and inviting public space – a park which knits into its surroundings through a clear and logical urban strategy'. Importantly, the plan included a housing development which was seen as integral to the park, recalling the strategies at Birkenhead and Sefton. The main concern of the London Development Agency, who have maintained the garden so far, is that Newham Council will not have sufficient resources to continue the maintenance at the same level.

The masterplan for the Dutch Floriade at Zoetermeer, 1992, which was in development for at least five years, was amongst the first designs to illustrate what the Dutch themselves called post-Romantic or post-Geometric design (the geometric in this case referring to the Grand Manner form). Nonetheless its patte-d'oie, triangle and circle forms were very geometric and its detail distinctly abstract. Since then, the next Floriade as well as several German garden shows all adopted the style. The Thames Barrier Park would probably not have raised as many eyebrows had it been a design for a site in mainland Europe. Is it then an abstract park or is it the park as art? It's hard to tell. It certainly challenges users to consider the park in a different light. The central, sunken zone is so unlike anything that the public expect of their open space and unsuited to normal park usage that it has the feeling of being a work of art, a Bridget Riley from the 1980s no less. It's the park as abstract art.

OTHER SITES

Under construction at Cramlington is Jencks' Northumberlandia, described as the world's largest human form sculpted into the landscape. It uses 1.5 million tonnes of soil from the Shotton Surface Mine and is due to be opened to the public from 2013. A five hundred metres long land art sculpture in the form of a reclining woman is just that. It may be wonderful and is undeniably bold. It is mentioned here for these virtues and because it will eventually will be part of an urban fringe park. However, it is likely to be best appreciated from a hot air balloon.

'Jupiter Artland', at the fabulous Jacobean (1622) Bonnington House, Edinburgh, is really a sculpture garden comprising commissioned pieces from leading sculptors and land artists and has changed as new pieces are added. In that sense, there is no master design, but this is reflected in the fact that there is no carefully devised route, nor even a network of footpaths. The garden opened in 2009 and contains contributions from both Jencks and Hamilton Finlay, amongst many. The facade of the house, very pleasing to the eye, might reasonably qualify as an exhibit in its own right.

Hamilton Finlay also contributed to an area of Stockwood Park, Luton, called the Improvement Garden. Like Jupiter Artland, it would be hard to classify it more as art than park because essentially it's a sculpture garden, rich in classical allusion, as might be expected, and containing six pieces set apart from each other with contrived vistas between them. Hamilton Finlay's contribution to the garden as art is really confined to his own garden, but this other work, modest as it is, is interesting in that it almost resembles pastiche.

The Laban Dance Centre in Greenwich, London (opened in 2002) has a turfed landform landscape which is composed of a collection of pyramids. In an attempt to conserve local populations of the black redstart, no trees were planted at the site because they would have allowed cover for predatory hawks. The landscape architects were the Swiss firm Vogt, to whom must go the best excuse ever offered for creating a garden without trees. Visually, trees would not complement the landform design but it is hard to believe that the pyramids derived directly from a wish to conserve black redstarts. Fifty-three years earlier, the Swiss landscape architect Ernst Cramer had made similar pyramids of grass at the G59 Exhibition in Zurich. Called the Garden of the Poets, it left the public confused and was a controversial component of the overall lakeside design. Similar effects were tried at the British Garden Festivals in the 1980s with less technical success.

Derek Jarman made a very fragile, unfenced garden on the shifting shingle of Dungeness, Kent, from 1986. Plants, driftwood and various found artefacts populate the area around the black shingle Prospect Cottage, apparently in a haphazard way. The landscape in which it sits is one of big sky, big beach, big sea, big power station, small and scattered dwellings. The garden doesn't interrupt the flow of any of this, but recalls the kind of games that children play when given a limited palette of found items, flotsam and sea-washed rounded cobbles. Stones with holes in sit on top of sticks planted vertically in the ground. Rusty bits of metal share the ground with lavenders and Californian poppies. To some it's art, to others not even artistic, merely a random collection of stuff.

Legacy

It's too soon to tell what the legacy of these gardens might be. The gardens of Jencks are certainly attractive, in that they attract a great interest from curious minds that have never seen anything like them. They invite exploration and cannot be appreciated simply by viewing from selected locations. To maintain their appeal involves a considerable investment in turf maintenance which cannot always be guaranteed. The same investment is required in keeping the intricate planting patterns he uses at Portrack in pristine condition. In a hundred years time will the elaborate stepped turf earthworks be lost under maturing trees, recalling Claremont? It would be a shame. However, these gardens can never be mainstream and should be viewed instead as artworks, the depth of their inspiration mattering much less to the public, who want enjoyment and photo opportunities, than it does to their creator.

Little Sparta and Prospect Cottage both demonstrate that it is perfectly possible to make gardens in extreme environmental conditions, without two spits of fibrous loam. Yet it was not the intention in either case to turn horticultural textbooks inside out but to draw inspiration from their locations and to see what might result. Without having first-hand knowledge, it appears from their form, character and locations that both might enjoy a modest maintenance regime. 'Low maintenance' has been a clarion call for designers for at least seventy years and is likely to remain so given that labour is never going to be cheap again.

9

TECHNOLOGY GARDENS

Since the Second World War, technology and circumstances have combined to provide the conditions for buildings and gardens to fuse in a way that, in extreme cases, it is difficult to determine which is the more dominant. Shortage of space in towns and cities has been one of the driving forces behind architects incorporating gardens within and on top of new buildings, typically offices but also public buildings, hotels and retail units. Roof gardens and interior gardens are included in these situations because they are seen as enhancing the environment and in some cases actually helping to improve air quality by stripping out noxious gases. They have developed as a response to a perceived need for better conditions, whether it be for working, shopping, eating or recreation, and where the economics of new projects was favourable there was something of a boom in providing both roof and interior gardens from the 1980s onwards.

Roof gardens

Although roof gardens have been made throughout the twentieth century, advances in green roof technology in Germany during the 1960s provided a new impetus for the movement which has now spread across the world.

Well-known gardens from the earlier part of the century are mostly in London. Selfridges store on Oxford Street had one in 1909 and Derry and Toms of Kensington High Street followed in 1938. Both were hugely popular with the public and became destinations in their own right. The Derry Gardens, as they are known, were designed by Ralph Hancock, who had created the rooftop garden at the Rockefeller Center in New York, but in London he had much more opportunity to use planting and water. There are still limes and oaks from the original planting, which is remarkable for two reasons. All the planting situations were given 900 mm of topsoil which is normally insufficient for stabilising taproots to properly develop.

The wind speeds at roof level are higher than at street level, so extra precautions should be taken to avoid the likelihood of uprooting but Hancock doesn't appear to have incorporated any guying. An innovating detail was that the water for the cascade and pool feature as well as the irrigation system was sourced from the company's own artesian wells. The Derry Gardens, which extend to an acre and a half, unusually incorporate spatial subdivision and the different areas were given individual treatments such as the Spanish Garden, the English Woodland Garden and even a walled Tudor Garden. They are listed Grade II and are still open to the public.

Geoffrey Jellicoe made a roof garden on Harvey's store in Guildford in 1958. There had been an exhibition on the work of Roberto Burle Marx at the Institute of Contemporary Arts earlier and its influence is plain to see in Jellicoe's design. Sadly, the scale was badly judged, as was the detailing. It was rebuilt in 2000 in the spirit of the original but on a smaller scale.

Selfridges reopened its roof garden in 2011 for the first time since 1940, but not in the same format. There is now a boating lake with twelve rowing boats and green water, all of which heavily surcharges the structure which itself has been reinforced with what the owners call 'miles of steel'. Whether it will remain for very long is a moot point as the new arrangement is called The Truvia Voyage of Discovery art installation, which is a phrase that has a sense of impermanence about it. Norman Foster's Willis Faber & Dumas Building in Ipswich, which was completed in 1975, has a very simple, level grassed garden surrounded by a hedge on the building's periphery. It really couldn't be any simpler and the space is interrupted only by the rooftop restaurant. The architects say that it offered a new social dimension and that the design 'was conceived in the spirit of democratising the workplace and engendering a sense of community'. As a minimalist design it offered no variety of experience for the office staff and in many ways represents a classic hostile environment. There was always a suspicion that having a green roof here was more about fitting with the clean lines of the building than it being of genuine utility.

Derek Lovejoy and Partners had a distinctive house style to their landscape design work from the company's establishment in 1959. This was recognisable in their roof garden designs as well as their other work and consisted of angular paving arrangements and precise construction reflecting the detailing of the Modernist buildings which often provided their context. The Halifax Building Society Headquarters, completed in 1974, has roof courts designed by Lovejoy which are enclosed by being made at the floor level of the top storey of the building. Reflecting pools are a significant element of the design, which always had minimal maintenance as one of its principles.

Dame Sylvia Crowe, working from London, made a roof garden over a multi-storey car park for a new Scottish Widows Life Assurance building in Edinburgh in 1975. She used a limited planting palette that included birch and sycamore with heather ground cover, in an attempt to 'merge the composition of the building and landscape together, to do justice to the very important site, both at close quarters and from the high ground of Arthur's Seat'. Regrettably, the building by Sir Basil Spence, Glover and Ferguson has an alien roof plan of interconnecting hexagons which allows no possibility of amelioration when seen from the high ground of the adjacent Queen's Park. Further, the choice of birch and heather has always seemed a pastiche of a Scottish wild landscape and perhaps would not be everybody's choice for a city centre project.

In Edinburgh, The Michael Laird Partnership built administrative offices for Standard Life Assurance Company in 1991. The building, now altered, had the largest roof garden in Europe and used lightweight materials to create elevation for mounded planting areas. Evergreen shrubs and trees in conjunction with a raised parapet helped to create sheltered zones for staff sitting areas. Integral to the building design concept was the advantage of insulation that the roof garden provided.

In 2011, the 60th anniversary of the Festival of Britain was marked by the creation of a roof garden on the Queen Elizabeth Hall, next to the Royal Festival Hall on London's South Bank. Unfortunately poor architecture cannot be redeemed by such a gesture.

In the USA there has been a movement towards using roof space to grow food. Chicago has led the way, having always used areas within public parks for food-growing projects, but now it has started to look at the 'wasted' space on top of its buildings, and set an example with the City Hall Rooftop Garden. Monitoring has shown a significant temperature reduction both at the surface and of the air, confirming that such gardens can contribute to the cooling of city centres as well as reducing energy usage. Sustain, an alliance for better food and farming representing public interest organisations, produced a report based on its findings in the USA which has led to increased interest in the creation of roof gardens for this purpose, specifically in London as part of the Climate Change Adaptation Plan.

There are two ways of establishing a roof garden. Either it can be designed from the beginning in which case the loading can be measured and allowed for in the building's structure, or it can be retro-fitted, where the structure may dictate exactly what can and cannot be achieved. The engineering as much as the architecture holds the key. Not only are wet soil, water, paving and trees heavy, the loading must allow for people and, depending upon the scale, machines. There has to be an appropriate access for garden materials, maintenance equipment and for the removal of waste. This usually entails a dedicated elevator system. Even when the roof garden is integral to the building's design, there will be an imperative to keep the weight of roof surcharging to a minimum so as not to unnecessarily affect the structure.

In most cases, a roof garden is likely to be technology driven and the designers must work within strict limitations set by the structure. They will only ever be

Landform endures

Particularly when covered in vegetation (grass is ideal), landform proves to be an enduring design tool. Early experiments, as here at Studley Royal, took geometric forms until the mid eighteenth century Landscape Improvers introduced naturalistic landform. Since the 1980s there has been a widespread return to the geometric and obviously artificial arrangement of earthworks.

Light through dark

The interest in any arrangement of spaces is greatly enhanced by there being a contrast of open areas and enclosed ones. This leads inevitably to light and dark zones, each having their own qualities which in turn affect the way people feel about them and use them. Bright areas of open space are inviting, particularly when seen from within a dark, shaded zone, as here in Sefton Park, Liverpool. No footpath leading from one to the other is required because the attraction is inherently strong.

In confined areas, the same effect can be used to encourage movement through a series of spaces. The garden walls at Little Thakeham, Sussex, create conditions of light and shade to frame views and emphasise the linkage of different areas whilst also reinforcing the sense of enclosure.

Lancelot Brown employed 'the line of beauty' where land meets water. The margins of his artificial water bodies or of modified river banks, as here at Audley End, Suffolk, were maintained not in a naturalistic condition but with clean, curvilinear lines which alluded to a perfection of nature.

Two gardens in the Grand Manner

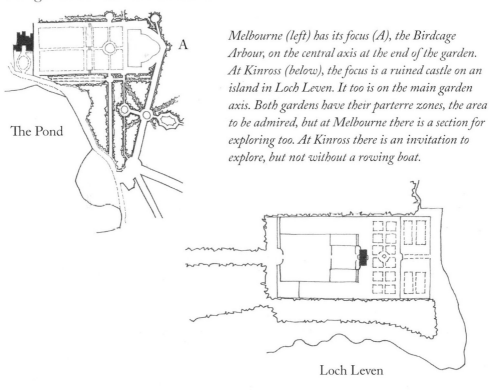

A

The Pond

Melbourne (left) has its focus (A), the Birdcage Arbour, on the central axis at the end of the garden. At Kinross (below), the focus is a ruined castle on an island in Loch Leven. It too is on the main garden axis. Both gardens have their parterre zones, the area to be admired, but at Melbourne there is a section for exploring too. At Kinross there is an invitation to explore, but not without a rowing boat.

Loch Leven

A

Mounts

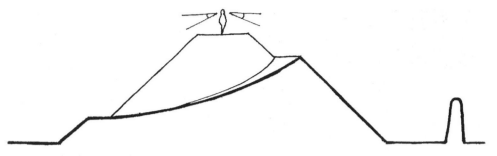

Mounts were usually positioned close to the garden boundary and afforded views both beyond the walls to the untamed landscape and of the parterre garden within. The grandest featured banqueting houses at their summits.

Munstead Wood, 1897(left) and Bentley Wood 1938

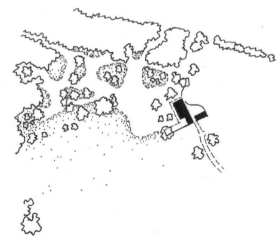

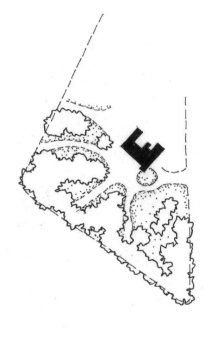

Both gardens were made with the benefit of existing wooded areas which were thinned to allow pathways to extend from lawn areas close to the house. Either side of the mown paths were bulbs and other planting set in long grass. The solution is suited to both architectural styles despite their being so completely different.

Outdoor corridors

Consistent with gardens being analogous to rooms, their design evolved, significantly in the Edwardian period, to include other architectural elements such as corridors. Pergolas are always more effective when they link other elements and have some kind of focus. Their definition can be determined with light structures that suggest enclosure of space and they work well when only partially planted. Indeed they are generally more satisfactory features than their predecessors, the Yew or Laburnum tunnels, being neither too oppressive on the one hand nor reliant upon a brief seasonal effect on the other.

Paving

White stone and red herringbone brickwork contrast with each other and offer a permanence of detail in a design which relied heavily upon herbaceous planting and was therefore mainly a summer effect. The permanence is dependent upon good foundations, otherwise expensive maintenance will be a recurring feature. The stone edging here originally separated the path from borders, allowing the planting to flop forward without causing any blockage of the path itself. Now, with grass replacing the borders, the design is much reduced in effect and unchanging through the year. The other image shows the designer's original intention of planting abundance contained by the geometric order of the paving.

The Picturesque

Cragside, Northumberland, is a late flowering of picturesque traditions that were introduced about a hundred years earier. In this exaggerated version of the scene at Castle Howard, every effort has been made to enhance the romanticism of the composition. Norman Shaw's faked aging of the house architecture employs Elizabethan chimney stacks and half timbered gables whilst the natural drama of the landform allows for a judiciously placed elegant footbridge to provide a middle distance focus and for both structures to be reflected in the water. Broad scale composition is paramount here. There is no garden in the traditional sense.

Restraint

Designers should always visit their sites several times and under different light conditions before settling upon a design. Nevile's Court, Trinity College, Cambridge, encloses only a lawn, but an exquisite one, particularly in the early morning light. Any planting, landform or water feature, although easily accommodated, would be totally inappropriate in this space, the lawn providing sufficient foil to the colonnaded courtyard. Landscape architects and plantsmen might disagree on this, but part of the skill of designers through the centuries has been an understanding about when to show restraint. The simple solution is often the most satisfying.

The simple arrangement of open, light space and closed dark space, the two linked by a pathway, is a device that, because of its simplicity, should always succeed, needing no signposts to encourage exploration.

A dark background, such as an evergreen hedge, can be a perfect foil to light foliage or flower colour.

Space linkage

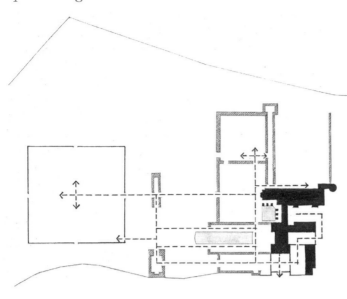

The garden space at Folly Farm is almost entirely geometric in composition and is subdivided into individual units. Close to the house, walls project from the building lines to create small courtyards. Spaces increase in size with movement through the sequence, hedges being used along with avenues to enclose and emphasise direction respectively. Even the large walled garden is aligned upon one of the principal views from the house and the carefully structured layout is both complemented and contrasted by an irregular area of lawns with trees and bulb planting on the west side (top) where the boundary is a stream. The order and progressive variation in scale exhibited here is not replicated in some of the more famous gardens of the early twentieth century, where horticultural exuberance is more to the fore.

The advantage of overview

The earliest geometric parterre gardens of the sixteenth and seventeenth centuries relied upon high level terraces and viewing points from within the houses for the appreciation of their detail. This was not lost on the classical revivalists of the mid- and late nineteenth century although what seemed to be more important, as here at Mellerstain, was that an axial view of the gardens took precedence over any utility or enjoyment that they might provide.

Lawn maintenance can be one of the most expensive of all costs in a park or garden but selective mowing can realise a relatively inexpensive approach to the organisation of access as well as providing a contrast of textural quality which is seductive and simple to achieve. This example is at Manderston House, Scottish Borders, but designers like Graeme Moore use similar techniques to achieve ground patterning for its own sake.

Water as reflector

A contrast of light and dark spaces adds interest and depth to most designs but it is sometimes good to bring light into the dark and nothing performs this better than water. At Tintinhull a block of stone has been carved out to form a shallow reflecting pool, no more than a bird bath in scale, but set on the ground under a dark spreading conifer it holds a mirror to the sky. The sinuous rill at Rousham has been copied many times but it remains the best example of its type. Winding through a dark tunnel of yews and cherry laurels, it shows the way from Venus's Vale to the Cold Bath and beyond in the manner of a shimmering ribbon.

successful if the environmental conditions are treated as a priority so that people can comfortably use them.

There have been instances where a roof garden has helped to secure planning approval for buildings in particularly sensitive locations, but they are anyway likely to become more frequent aspects of dense developments in city centres. In 2008, the World Green Roof Congress was held in London and used the geometrically designed roof gardens at Bishops Square as a demonstration venue.

Interior gardens

From the late 1970s, interior landscapes have flourished. At least, as a type, they have been a component of very many buildings but they have not always worked as well as they should have done. This is usually because the specific environmental conditions for healthy plant growth differ from those which suit office workers sitting at computers, or hotel lobbies where doors are constantly opening and closing. Like roof gardens interior gardens will only be successful if the architects and the garden designers formulate their plans together from the outset. In 1985 I was asked to address the British Association of Landscape Industries, Interior Landscape Group Conference in Edinburgh, on the subject of architecture working in conjunction with interior landscapers. Referring to my own experience up to that point, I explained something of the frustration involved with designing interior installations and how on one project the other designers were only interested in getting it all looking right on the night – not the opening night, but for the photo session for *Domus* and *Design* magazine. 'If the plants die, well, we'll just get some more,' was the response to my protestations. Fortunately not all projects are like this, but many were. I remember that not everybody in the conference hall seemed shocked.

Light, temperature and irrigation must all be considered and allowed for adequately. Of all these, my own experience suggests that light is the most critical and probably the most difficult to accommodate. Failures of planting are often traceable to lights being switched off, or to light bulbs giving the incorrect spectrum for plant survivability, or simply to insufficient light being allowed for within the design. Plant requirements are known and both light fittings and glazing have measurable qualities so there is little excuse for failure. During the 1980s, when the fashion for interior gardens made rapid headway, the *Architects' Journal* helped to fuel the myth that anyone could quickly become an expert in how to design interior landscapes. The journal offered pull-out sheets presenting technical information and lists of suitable plants in a way that trivialised the process of design. The implication was clear that by following a set of easy steps the architect could dispense with the services of other professionals. There were architects who believed the myth and there may still be.

Temperate zone plants are generally unsuitable for interiors because buildings are maintained at even temperatures throughout the year. It follows that tropical and sub-tropical plants are better suited. Specialist nurseries in Belgium, Holland and Florida have been the usual source of large nursery stock. In Florida it is grown outside or under extensive top-sheltered, open-sided frames.

Getting the design right is not easy when working in entirely artificial circumstances. Specialist advice should always be sought because small mistakes can be catastrophic and, of course, very expensive.

Given that the design parameters are met, plants can be very successfully grown inside. Typically architects arrange deep-plan offices and hotels around atria so that edge situations, known to offer preferred working conditions, are available to more people. Coutts Bank in London has a well-designed interior garden by Frederick Gibberd and Partners in its bright garden court, otherwise known as the banking hall. Polished stone, leather seating and healthy planting make a relaxed and luxurious waiting area which looks as fresh now as when it was finished in 1969. Gibberd had originally wanted to use temperate zone plants but quickly realised that this wouldn't work in an inside environment. Not wanting to have the aesthetic of a tropical jungle, they chose instead sub-tropical plants, including *Magnolia grandiflora*, *Acacia dealbata* and several small-leaved *Ficus* as the tree components.

At the Grafton Shopping Centre in Cambridge, constructed in the early 1980s with landscape design by the Derek Lovejoy Partnership, an atrium planting scheme had hanging planters at a high level. Some daylight was available and was supplemented by artificial lighting. One problem with this type of scheme is that looking up one sees the underside of leaves and the blinding light of halogen bulbs directed downwards from the roof. Achieving the correct orientation of the viewer, the planting display and the light source is vital to a successful scheme. In 1982 Maurice Pickering Associates won the British Association of Landscape Industries (BALI) Grand Award for their interior scheme at the AMRO Bank in London EC2. It had a pool and cascade on the ground floor of a seven-storey atrium. The water produced humidity for the atrium planting and helped to cool the building by about 3°C. Water has perhaps been underused as a feature of interior designs, but Pickering made the point, soon after the building's completion, that the cost of interior gardens is minute compared with the general building costs and their expense is quickly recovered in speculative projects like this by charging higher rents. He also suggested (*Landscape Design*, April 1983) that staff were more contented and productive, although without offering any support for that view. The Milton Keynes Shopping Centre, now listed Grade II, was completed in 1979 and set a design standard for retail centres that has not been matched since. Plenty of natural light, innovative solutions with planting and high-quality finishes like Travertine marble justify the listing. Two arcades run parallel with each other, the south-east facing one having plants from hotter and drier climates while the north-west one has more of a tropical forest character. The cost of the plants and their planting was £100,000, representing less than half a per cent of the building costs. Only a year after it was opened, botany lecturers were taking students there on educational field trips.

Gardens in and on buildings have evolved quickly and are no longer considered to be 'cutting edge' simply by having parapets and balconies dripping with trailing plants – their purpose to decorate the architecture. Unfortunately, interior gardens are still planned using synthetic plants because some clients put more stock in illusion than reality. Plants can enhance the interior of buildings at the same time as contributing to air cleansing. Insulating buildings is just one of the benefits that roof gardens can offer. For these practical advantages to work it is more imperative than in any other situation that all the designers are working together from the outset. Failure is not an option and techniques must be soundly understood, being built in to the creative process.

10

ANOMALIES

Some gardens don't fit anywhere conveniently. In another fifty or hundred years it might, indeed should, be easier to recognise trends and influences which are currently hard to find. Overtly arty gardens may become the norm, although I doubt it. It is perhaps more likely that people will want their own gardens to deliver the traditional properties, not least that they will be comfortable living areas outside and related to their homes. Anything that is unusual will attract the attention of the curious but that doesn't mean that overtly artistic or abstract gardens will ever be anything more than exhibits to be visited, at the same time dividing opinion in the way that the Turner Prize for new developments in contemporary art always attracts as much negative as positive criticism.

Sir Geoffrey Jellicoe designed a number of gardens during the twentieth century which sometimes embraced Modernism but more often didn't. One of his biggest commissions, Sutton Place, came late in his life and his solution for it was to revisit history, a decision which might have demonstrated his respect for the old house more than a wish to be creative with a contemporary garden.

One of the biggest garden attractions of the twenty-first century is the Eden Project in Cornwall. Cruise liners include it in their itineraries and have helped to boost the visitor numbers to eight million in the first ten years. In some ways this could be categorised as a technology garden, but it is much more than that and so far is unique in Britain.

Whatever the composition or content, it is hard to find any aspect of these sites which takes the process of garden design into unchartered waters.

SUTTON PLACE

Sir Geoffrey Jellicoe
1980s

Sutton Place, near Guildford in Surrey, is a curious, unique garden with design elements by Geoffrey Jellicoe in the 1980s and plans of earlier work by Jekyll from about 1900, which may or may not have been carried out. The house is early Tudor, started in about 1525. Apart from incorporating some features into existing enclosures, Jellicoe saw his design for a serpentine lake to the north of the house realised but not the cascade which was planned for the southern prospect from the house. An orchard has been planted here under more recent plans.

Jellicoe was a giant in the profession of landscape architecture during the twentieth century and he wanted to make a garden which, in his own words, 'expresses the modern mind, was sympathetic to the ethos of the place, which comprehends the past, the present and the future'. Whatever this means, early access to the garden, in the late 1980s, was only allowed under the guidance of an estate courier who was charged with explaining the garden to visitors. There was an air of the precious about the new additions, which include a Paradise Garden, a Moss Garden, and a Swimming Pool Garden (with a Miro-inspired raft). Some of the garden ornaments from the sale of Mentmore Towers have found their way here and there is also a white marble 'wall' by Ben Nicholson, placed for reflection in a pool. Whatever the quality of the individual gardens, and there is no denying that the workmanship is of the best, this is a strange mix of Tudor enclosures, Victorian-like eclecticism, twentieth-century plantsmanship and an eighteenth-century style lake of a form which might have been drawn by a lesser exponent of The English Landscape Garden. In some ways it can be compared to Highgrove, but without the farmyard manure. Equally it would be reasonable to put it in the same file as Chatsworth and others like it for its eclecticism gathered over time. It might ultimately be nothing more complicated than a sculpture garden but at the time of writing the received wisdom is that it might be something more.

The requirement for a courier or guide ensured that the allegory and analogy, with which the design is heavily imbued, was made clear to all visitors. By implication, Jellicoe and his client were unhappy about anyone merely enjoying themselves in the gardens. There may be a considerable body of people who believe that whatever the inspiration might have been for a garden, it should primarily be a place of enjoyment, whereas it is perfectly reasonable for the deep meaning of life, the universe and everything to be pondered at leisure by staying in bed and reading Douglas Adams.

The Eden Project

Tim Smit, Land Use Consultants, Nicholas Grimshaw, Anthony Hunt & Associates
2001

Apparently the Eden Project has the largest greenhouse in the world, which statistic is probably enough to attract interest on its own. Parts of the 'garden' are contained in bubble-shaped enclosures which occupy a disused china clay pit near

St Austell and, as a project, it is still developing. The bubbles are geodesic domed tubular steel, space-frames with panels or cells of thermoplastic ETFE (ethylene tetrafluoroethylene) which is UV-transparent. This material has also been used to clad the visitor centre pavilion at the Alnwick Garden. It has amazing properties, being one per cent of the weight of glass, transmits more light than glass and costs considerably less to install. In the very unlikely event that it catches fire, the consequence would most certainly spoil your day out, as highly corrosive hydrofluoric acid would rain down from above.

The project reproduces biomes or ecosystems, two of which (tropical and Mediterranean) are in the covered domes and one (temperate) is outside. In this respect Eden's covered areas differ from glasshouses in botanical gardens by incorporating fauna as well as flora. The most recent addition to the project is an environmental education facility called The Core. The roof of The Core has been inspired by phyllotaxis (specifically the opposing spirals found in plant systems) and has a photovoltaic array, less than half of which receives direct sunlight. This has understandably been the source of some controversy.

The interior biomes are visually dominated by the domes, which have the same effect on the landscape outside, albeit contained within the landform of the pit. Their bright, white and reflective bubble-like forms sit unhappily in the context of an enfolding topography, apologetically hiding from the wider landscape. Car and coach parking is on a necessarily massive scale and is arranged in long ranks, slightly curved. This helps to visually absorb the vast numbers of vehicles.

Britain's first Biome Garden is no Eden. It does have green credentials but not four rivers and an apple orchard. There is an artificial temperate zone ecosystem that sits between the car parks and the giant domes and comprises 'native plants and plants from the temperate parts of America, Russia and Indian foothills that share a similar climate to our own'. Whether this can reasonably be classified as a biome is open to question and some of the project's literature only refers to it as 'The Outdoor Landscape'.

Smit was also responsible for The Lost Gardens of Heligan, a restoration project in nearby Mevagissey, which is discussed in Chapter 13.

OTHER SITES

Jellicoe also designed a garden at Shute House, Wiltshire, in 1978, adapting an older plan by inserting geometric flower beds and a rill cascade which was supposed to create harmonious sounds. The character of the garden generally is a blend of cottage garden meets William Kent and, as with all Jellicoe's work, is rich in symbolism and derives from his deeply held belief in design being married to the subconscious. He strongly admired the work of Henry Moore, Barbara Hepworth and Ben Nicholson, finding room for their pieces whenever possible. Art critics, writers and observers are now starting to question the real worth of twentieth-century art icons (Laura Cumming, *The Observer*, 28 February 2010 on Henry

Moore at Tate Britain: 'Moore made far too much to be consistently original. His work became as repetitive as it has become familiar'; Jonathan Jones, *The Guardian*, 23 May 2007: 'Gormley is the new Henry Moore: third rate'). Jellicoe's own work, which has been revered for so long, could very reasonably be reassessed as muddled, otherwise known by some as post-modern (but see Chapter 8). Unusual amongst landscape architects of his generation, Jellicoe had little knowledge of plants and trees, admitting to an audience of students in Atlanta, Georgia, that he had no idea what trees he would be specifying for his project at the Moody Gardens, explaining that he had never been able to distinguish one tree from the next. For him landscape design was an expression of balance and proportion, of light and shade and of rest and movement. It is all of that, but Lord Lindsay was in Jellicoe's party on that visit and he remembers Jellicoe concluding the discussion with this: 'As for the difference between a pine and an oak – that is a mere technical detail' (*Landscape Design*, September 1996).

Landscape architects Colvin and Moggridge are responsible for a new garden at Williamstrip Hall, Coln St Aldwyns in Gloucestershire. The geometry of their design superficially resembles the east front of Hampton Court Palace with a patte d'oie of avenues and linked circular parterres close to the house. True, the canal and conical yews are missing, but there is still room. The design style predates the house in this instance. In the twenty-first century it is rare to find a client willing to invest in such grand landscape planning and presumably have the confidence that its maintenance will be ensured. It is also unusual to find a designer willing to turn the clock back 350 years without making at least a modest contemporary statement. It would be strange indeed if Sutton Place and Williamstrip Hall became trendsetters for today's garden designers.

The Alnwick Garden is a modern development in the grounds of Alnwick Castle, Northumbria. It was made for the Duchess of Northumberland by Belgian designers Jacques and Peter Wirtz, and opened in 2001. Described in publicity material as a 'garden for gardeners and for families' and 'a garden that looks to the future', it is an eclectic collection of styles and details that loosely resembles the Garden Festival approach to design (see Chapter 11) but the quality of workmanship is rather better. That said, the Italian Renaissance-inspired cascade has unconvincing detailing and its orientation to be in the shade strikes a false note. So often through garden history, water features have been designed on a grand scale at great expense but the smaller, more subtle treatments inevitably provide a more satisfying result. The narrow rill that winds its way through an evergreen tunnel at Rousham has so much more appeal than a cascade that faces the wrong way and shows its full hand in a single view. The deep shade is also entirely appropriate there. The Wirtz family love elaborate and extensive evergreen hedging, a feature which characterises the great majority of their gardens including this one and which imposes a maintenance burden at least as onerous as that the late seventeenth- and early eighteenth-century landowners faced in keeping their estates in good order.

Legacy

What has become clear is that at the beginning of the twenty-first century there is still a strong vein of eclecticism pumping through the garden design world in Britain. The Victorians liked to have a bit of everything within their gardens, but this is different. At no time in history has there been such a diverse range of styles being produced simultaneously, along with the background of historical restorations and reconstructions which are continuing with vigour. No longer are designers pontificating on how gardens should be designed or arguing amongst themselves about the rights and wrongs of particular approaches, because today anything goes. Only when there is a possibility of heritage being tampered with do people tend to get exercised about the details.

Some designers, supported by observers, think they might have created a new way of designing under the banner of Conceptual Gardens. Again, *The Telegraph*'s Tim Richardson is in the vanguard of categorising these works and he even offers a shortlist on how to become a conceptualist gardener, including such advice as: 'Every garden is a political act and every plant is a political prisoner; A garden does not have to be a retreat. It can also be an attack; Without meaning, a garden is just a pile of materials; Give the garden a theme and stick to it'. This last point is a very appropriate segue into the next chapter.

11

TEMPORARY GARDENS

Many people are quite content to have lawns with island rose beds and some area for bedding out summer displays of half-hardy annuals. This sentence could have been written 150 years ago and still be accurate. For those who have a little more ambition but are perhaps short on invention, examples of what might be possible abound, thanks to the annual shows at Chelsea and elsewhere. Sharing the stage also, for a brief period, were the Garden Festivals. Necessarily, these collections of ideas are, or were, assemblies of small themed gardens, self contained and intended to spark ideas in the public's imagination. It might be disingenuous to suggest that some are intended to win medals first and spark interest second.

GARDEN FESTIVALS

Germany and Holland had established a good tradition of Garden Festivals in the years following the Second World War, the German plan being to hold them every second year and the Dutch, rather conservatively, every tenth year. Hanover hosted the first one in 1951 then all the major German cities took their turn, with an international event being awarded to the country every tenth year.

The German festivals, *Bündesgartenschauen*, were a response to their bombed-out city centres, the festivals reclaiming land and leaving city parks as their legacy. The lead time for these sites was often a full ten years although sometimes only half of that was the construction phase. The results, at least up to 1991 – my first-hand experience stopped at this point – were generally very positive. When they opened their gates the parks had a respectably mature framework of structure planting. In some cases it was more than this because sites were reused (Hamburg, Essen, Dortmund) in later years. The standard of design was robust, sometimes innovative, always attractive.

In Holland, all the festivals – known as *Floriade* – were, indeed are, of international status and again the sites remained as parks after the summer of each of the festivals. Long lead-in times are recognised as important to their success.

Between 1984 and 1992 Britain joined in, with five Garden Festivals at two-year intervals. Born from a political initiative that had job creation, reclamation and regeneration in mind, the government part sponsored the programme of festivals and selected sites that were most in need of recovery, in each case within a different region. The first site was on the banks of the Mersey in Liverpool and attracted international status. All eyes were on the British landscape profession to see whether it could rise to the occasion with only a short lead time, of about three years in the earliest cases. All the chosen sites were in areas where a determined industrial and/or housing after-use was agreed because the government had already concluded that the country had no further need for urban parks. Only a token open-space legacy was allowed for, so the landscape finishing works were largely seen as a temporary surface dressing to reclaimed land and would be dug up after the five or six months of each festival.

Although not principally garden orientated, there have been other temporary festivals and exhibitions which have had some effect upon landscape and garden design. The Great Exhibition of 1851 showcased above all else the pavilion, Paxton's Crystal Palace. When the exhibition closed, Paxton bought the great glasshouse and moved it to Sydenham Park in Surrey, where it eventually burnt down, but not before having captured the imagination of gardeners and architects alike. The 1951 Festival of Britain left us with one fine concert hall which was eventually surrounded by unlovable concrete-dominated buildings and graffiti-covered undercrofts known as the South Bank. Next to it the Jubilee Gardens of 1977 have been an empty, open grassed space of poor quality for too many years, the original layout having been unworthy of retention after the year's celebrations. The Battersea Park Pleasure Gardens, designed for the 1951 Exhibition, seem to have been something conceptually approaching a Garden Festival of the Thatcher era but set in a mature park landscape. Descriptions and illustrations of it in *Design* magazine, January 1970, show that it was a glorious exercise in ghastly kitsch with sprinklings of Peter Shepheard's vaguely Modernist planting.

INTERNATIONAL GARDEN FESTIVAL (IGF) LIVERPOOL 84

Three and a half million visitors came through the gates. A comparative handful would have been landscape professionals, the others having been seduced by promises of a great day out. It's not possible to tell whether they were happy or not and in what way the design of the festival site affected them. There was large-scale ground modelling which helped with the macro spatial subdivision, but otherwise planting was relied upon to do the job and much of it didn't work. The mounding was well drained, there was exposure to winds off the sea and there was a drought in the spring of 1983. Large and small nursery stock suffered and had no time to recover. Much of it died. Even if the seasons had been kind, success would have been relative and the dark green shelter planting marked on the design plans would never have achieved sufficient maturity to be effective.

A very strange crescent-shaped banking, stocked with bedding plants and tulips and called the Victorian Garden, was like no garden any Victorian would recognise.

National Garden Festival (NGF) Stoke 86

Visitor numbers were down for Stoke's summer in the spotlight. Two million tickets were sold. My notes from the time tell me that it rained on both days of every weekend that summer, so the weather was never going to encourage people to go out and enjoy themselves. The masterplan didn't help. The Stoke site had a central ridge which was supposed to be 'wooded' but again there was insufficient establishment time and the reality delivered a bare hillside covered with windswept whips. The ridge separated the site, when unifying it would have been desirable. Mannheim's Luisenpark (1975) had a two hundred metres wide centrally located open space and a skypointer tower as a constant reference point. It worked very well. Views at Stoke were, because of the topography, largely off site and ugly. The central ridge idea was probably unique in planning terms. Most festival sites at home and abroad follow a pattern or formula which superficially resembles an eighteenth-century English landscape park with water bodies near the centre. Circuitous paths give access to the theme gardens and high-activity areas while the general orientation is inward looking, aided by peripheral planting. At Stoke the water features were peripheral: the canal was mostly invisible, the canal basin was bleak and located at the south-west corner where there was no encouragement to walk around it and no need to either. Some small lakes near the festival hall were not integrated well with the rest of the design.

This was a poor choice of site and the masterplan didn't manage to overcome the inherent problems. A cable car to the ridge top only exposed visitors to a wider view of dereliction, stronger winds and no shelter from the endless rain. Worse than all this, if that is possible, was the construction detailing, so often badly designed as well as badly put together, so simple to deconstruct with a well-aimed kick.

NGF Glasgow 88

The Glasgow masterplan team consulted Disney sites in America to help them understand the design issues of handling large numbers of visitors, so it was no surprise to see that a version of High Street America was included as the circulation hub. The focus of the site was not a high-quality piece of landscape design but an area called the 'central milling space' to which it was necessary to return time after time. Glasgow 88 relied upon landform, on a small scale, to separate the themed areas and it worked, at least partially, but there were so many paths that it was easy to get confused about exactly where you were. 'You are here' noticeboards were scattered at intervals to help with this problem, but simply orientating was a challenge. Notwithstanding this, four and a half million people made it through the turnstiles. The *Coca-Cola Roller* thrill ride vied for attention with the gardens and arguably performed rather better than most.

The Glasgow site was wrapped around a disused dock on the River Clyde. There was a new bridge, a tall ship and an old paddle steamer. There was also a simulated mountain stream and a garden where the pines had suffered so much in the lead-in to the opening that, for appearances, they were spray-painted green the day before the festival opened. The agenda was becoming clear. German quality landscape design was not possible in the time available, but 'giving the punters what they want: fun!' could still be aimed at. That quotation came from one of the landscape professionals involved with Glasgow 88.

Immediately following the festival, the Scottish branch of the Landscape Institute held a debate in Govan Town Hall. The motion was: 'That Garden Festivals in Britain have failed to achieve their landscape potential'. One well-respected delegate suggested that they were 'intellectually barren and moribund' and had turned into 'competitions to see who could build the biggest teapot'. In general though, those involved with the design tended to defend it while others were deeply worried about the profession getting itself involved with what was already recognised as a poisoned chalice. The design team had become reconciled that their efforts did not involve good and robust design. Onlookers wanted them to settle for nothing less.

Financially, this festival made a multi-million pound loss on the gates and it was known that this would be the case before it ever opened. Justification was made by claiming that invisibles would make good the shortfall.

NGF Gateshead 90

The Gateshead site attracted over three million visitors. To give a flavour of the event, this is from the *Gateshead Times*:

> Attractions included public art displays, a Ferris wheel, and dance, music, theatre and sporting events. Several modes of transport were provided around the site: a monorail which ran between Norwood and Eslington, a narrow gauge steam railway between Dunston and Redheugh, and a road train which covered the entire site. A ferry across the River Tyne, between Dunston Staiths and Newcastle Quayside, was also provided. The festival site was created over a two year period, on 200 acres (0.81 km²) of derelict land, previously the site of a gasworks, a coal depot and a coking plant.

Alert readers will notice that there is no mention of gardens, possibly because there weren't many and they were not the main attraction anyway. As a visitor attraction, Gateshead 90 fell well short of being a satisfactory experience. Having made an effort to travel in the expectation of spending a full day there, I was heading home inside three hours. There was no delight, no cohesion, no sensibility to anything. It was a low point, in the company of low standards.

NGF Ebbw Vale 92

By the time this festival appeared, this author had completely lost both interest and heart. He didn't visit. Despite fervent requests for more of the same from the interested professional bodies, the government decided to pull the plug, ending the Garden Festival era ten years after work started on the Liverpool site.

Legacy

The profession was on show and it found itself wanting. There were problems with masterplanning, detailing and plant establishment, not all the fault of the designers but they might have predicted them. Physically, there is precious little left of the landscape finishes, a good outcome maybe. The main legacy is that derelict city centre sites were reclaimed. The collective dream of British landscape architects who had been regular visitors to the continental festivals, of 'Now it's our turn', soon turned to a 'Get me out of here' nightmare of flimsily constructed pastiche theme gardens. The festivals only claim a place here because history should not be air-brushed away when it seems convenient to do so.

Peter Bareham, writing in *Landscape Design*, February 1983, gives an excellent, in-depth review of the Bündesgartenschauen, highlighting the role of the design professionals and the after-use of the park sites. It is an interesting counterpoint to the British experience.

FLOWER SHOWS

Flower shows have for many years included examples of what are variously known as show or display gardens to suit different circumstances. Not all flower shows do this as it takes a bigger site and a good deal more disturbance to stage show gardens than to erect marquees filled with floral displays.

The Chelsea Flower Show

The Chelsea Flower Show started life as the Great Spring Show in 1862, although the Royal Horticultural Society had held other shows since 1833 at Chiswick. It moved to the grounds of the Royal Hospital, Chelsea in 1913 and has recently extended its opening time from four to five days, always towards the end of May. Apart from staging the country's best, some say the world's best, horticultural show, the event also includes gardens categorised under several different types. Competition for medals in both horticulture and garden design is very fierce, success here being seen as a springboard to commercial fortune, or at least a big break.

The standard of workmanship, sometimes also of design, is usually very high and the show attracts designers who have something new to say, if occasionally in an avant-garde way. This has led to the comparison of Chelsea with fashion shows, although the publicity is much greater than, for instance, London Fashion Week, there being extensive prime-time television coverage nightly throughout the show. Space is always limited and tickets are not cheap. In recent decades the show has become part of the social scene and visitors can expect to see as many celebrities in attendance as there are chrysanthemums.

Feelings run high amongst the public. In 1993 a 'seaside garden' won the Wilkinson Sword Award for Best Garden but visitors to the show complained that a sand dune wasn't a garden. The year 2009 saw *Paradise in Plasticine*, a garden made entirely of a soft, putty-like modelling material. Most people enjoyed the humour.

Rarely, a garden is dismantled and rebuilt at the home of a visitor, who clearly wants a design original and is prepared to pay for it. Whilst the show recognises its role as a shop window, it can hardly have expected to be selling complete gardens and, by so doing, making the temporary permanent.

Arguably more than any of the 'outdoor room' type of plantsman's gardens of the early twentieth century, those on show at Chelsea send people away with ideas aplenty for improving or changing what they have at home because there is more variety and invention in one show than in Hidcote, Sissinghurst and Tintinhull rolled in together.

HAMPTON COURT PALACE FLOWER SHOW

Here it is again. We started this trawl through history in the Tudor sixteenth century, highlighting Hampton Court as the first case study, and it seems that there have been design developments there ever since.

If Chelsea is the best, this show, held at the beginning of July each year, is the biggest. Shows began in 1990 and the Royal Horticultural Society (RHS) began their involvement with it two years later. Like Chelsea, there are show gardens as well as the horticultural displays, and the whole event takes place either side of the Long Water in the park beyond the Great Fountain Garden. It has proved to be hugely popular and the location has been able to absorb the thousands of visitors quite effortlessly.

TATTON PARK FLOWER SHOW

The Royal Horticultural Society staged its first flower show at Tatton Park in 1999 and it returns every July, offering the same kind of experience that Londoners have grown used to, but in the north-west of England. There are Show Gardens and Visionary Gardens, the idea of the latter type being to challenge the way that gardens are made and even what they should be. Designers take this as an opportunity to bring out the inner artist and play with landscape materials without the worry of such things as durability and maintenance. In every case there is an underlying message, an allegory which is sometimes very deep and needs interpretation.

GARDENING SCOTLAND

The Royal Horticultural Society's policy of staging regional shows started in Glasgow in 1997. The chosen site was the island in Strathclyde Park's lake, but the target number of visitors was not reached in the first three years so the RHS withdrew, citing a financial shortfall. Gardening Scotland is the new name of the show and the Ingliston Show Ground, Edinburgh the new location. There were fourteen Show Gardens in 2011 and the visitor target was lower than anything the RHS managed to achieve in Glasgow.

Legacy

It almost seems that the plethora of show gardens have as their raison d'être change for change's sake, as if a little lateral thinking is always a virtue. If a groundswell eventually develops that achieves what William Robinson failed to do, then it will all surely have been worth it. However, creativity is a rare attribute and it always costs money. Prolonged economic recession focuses the mind and garden design will never make it to the top of a list of priorities that includes mortgages, food, fuel and football. However, for the designers, who are often sponsored, it must be huge fun creating a little paradise and being able to show it off to thousands of people. Excellence is always aimed for and sometimes achieved. The problem arises when inspiration turns to DIY and the materials, construction and site planning perhaps fail to produce the standard of the examples in the shows.

12

ESSENTIALS OF A GRAND TOUR

If the reader cannot manage to programme a tour of all the foregoing examples, which would anyway involve a truly heroic effort, there follows a shortlist of sites that might be considered as essential. The list is not quite in the 'See Venice and Die' spirit, not just because the original quotation might have been referring to Naples but because nothing can be that serious. Not all of these gardens are great, although together they comprise a representative cross section through five hundred years. Making the selection has not been difficult – it has been very difficult. The problem has been not what to include but what to exclude, although there are no flower shows or Garden Festivals for obvious reasons. Roof gardens and interiors are such specialist types that relate to particular circumstances and are therefore inappropriate to be designated as examples because even where they are creative, they are not typical.

Where to start? The obvious place would seem to be **Hampton Court Palace** for its restored Privy Garden and Tijou's screen to the Thames. Whilst there, the compromised Great Fountain Court could be viewed quickly, to properly appreciate why it doesn't work very well. The maze might be tackled too, but this is not the best maze in the country. A tour should include Oxford for the **New College** mount rather than the Botanic Garden, but why not visit both? The courtyards of most of the old colleges are also worth a look. **Edzell Castle** is included as the best example of a walled pleasance to a private home. Never mind the lawns and topiary, focus on the walls. **Kinross House** is here too because of its all-round quality and the off-site focus. **Bramham Park** chooses itself, being the best large-scale French Renaissance-style garden almost completely in its original form. For its approach avenues, early terrace, spectacular buildings, sublime vista and appropriately understated parterre, **Castle Howard** claims a place. No tour can be complete without **Stowe**, specifically for the Elysian Fields but also for the vista from the south portico. Arguably Brown's best effort with water, the lake at **Blenheim Palace**, deserves inclusion, as does the nearby **Rousham** for its exquisite garden, contemporary with the Elysian Fields but on a smaller scale. If only one of these two is possible, the choice must be made on personal preference: will there be more pleasure and instruction gained from the 'secret' small-scale, self-contained garden or from the linear garden of completely axis-free linkage? Choose either Stourhead or **Painshill Park** for an example of the eighteenth-century garden with multiple contrived prospects. The latter would get my vote. **Sefton Park** represents a mature, large-scale public park, which is where the Victorians were at their best. **Cragside** squeezes in for combining a picturesque setting of a woodland garden with an early Arts and Crafts house. The hillside garden at **Bodnant Hall** also deserves a place for its acceptable face of Italianate terracing. **Folly Farm** edges the choice from the Edwardian period as **Sissinghurst** does for its balance and controlled handling of effects as a plantsman's garden. The **Scottish National Gallery of Modern Art** seems an entirely appropriate location for the garden itself to be art, so this is preferred over any of the other landform sites.

I struggled over whether to include Studley Royal but decided against it mainly because of the new site arrangement and recommendations to visitors which are discussed in detail in Chapter 13. Somehow, it would be good to arrange the route so that a drive-past of Audley End could be included, perhaps on the way to Cambridge, but of course the Backs are more civic design than garden and, beautiful as the west side of Audley End is, it would be hard to justify it as one of the essential sites. So, my list is cut short at sixteen, averaging a fraction over three per hundred years and including examples of most of the identifiable types where there has been conscious design of spatial variety to create areas for specific uses or moods. This has therefore excluded botanical and national collections, arboreta and rhododendron gardens, in which the planning is not so much for people as for plants, so is more scientific than creative.

13

PARADISE LOST?

Throughout the text of this book reference has been made to gardens which have been restored or where restoration is in progress. It is a subject that exercises the minds of garden historians for good reasons, not least of which is that it's usually a very expensive process and should therefore be done accurately and properly. The definition of what constitutes an accurate restoration is where there has been much disagreement, not so much in the detailing of the design but in relation to the fact that, where a garden has been in development over a period of many years, choosing the appropriate elements or timeframe for restoration can, perversely, result in a certain amount of destruction.

In 1977 there was a sale of the contents of Paxton's Mentmore Towers, in Buckinghamshire, including all the garden ornaments and statuary. Although the garden was never of any great design interest, its virtual loss, along with all the furniture from the house, was a wake-up call for conservationists, fuelling an urgency to save the country's heritage. Jenifer White, Senior Landscape Advisor at English Heritage in 2007, explained (*Landscape*, May 2007): 'The job of English Heritage is to hand designed landscapes over to the next generation'. She sees their work not as preservation but conservation, which she defined as 'the process of managing change in ways that will best sustain the heritage values of a significant place in its setting, while recognising opportunities to reveal or reinforce those values for present and future generations. Conservation is a dynamic concept.'

A. A. Tait, Professor of the History of Art at the University of Glasgow, writing in the magazine *European Gardens*, summer 1996, suggested that respect for its past should not fix a garden for ever at the period of its creation. He says: 'I feel that if garden history could be forgotten and the doctrine of historicism abandoned, many critical problems might disappear. For behind much of this thinking is the wish to treat the garden as though it were a house.' He continues:

Gardens are assessed and rated according to their rarity within a cultural history stretching from the sixteenth century to the near present. Once identified they may only develop within particular stylistic categories. Gardens have become like salmon in glass cases, permanently fixed, frozen in motion, lifelike – but completely dead.

Gardens cannot be both fixed in time and changing with the needs, perceptions and value judgments of successive generations. Even small changes lead to more of the same and ultimately to the artistic value of the original being lost. Of course, the quality might even be enhanced by changes, so policies which allow no flexibility might be considered unsafe. Where should the line be drawn? Perhaps the principle should be that the essential qualities of a heritage garden are not lost by any changes that are made. Crucial to this is an understanding of both the original concept and motives of those who made the garden. John Sales, among many with the same opinion, has stated that it is the differences between gardens and not their stylistic similarities that ought to be concentrated upon when considering conservation. In his paper at the Conservation of Historic Gardens Symposium in May 1984, he pointed to the fact that much of gardening is accidental and that one of the traditions of English gardening lies in recognising the happy accident and knowing when to retain it. This is true, but of course gardening is not garden design.

A number of factors must be considered in the restoration and management of gardens and parks. The owners may be private, corporate, national or charitable organisations, each having its own priorities as well as the will to achieve the work. In addition there are separate registers of parks and gardens for England, Wales and Scotland. These give detailed descriptions and provide certain legal restrictions on any development which might affect them.

Registers

REGISTER OF HISTORIC PARKS AND GARDENS (ENGLAND)

The Register of Historic Parks and Gardens of special historic interest in England is under the control of English Heritage and was established in 1983 under the National Heritage Act. Any park or garden registered under this Act enjoys legal protection in that registration is a material consideration in the planning process. Planning consent must be given for any changes that are proposed.

The grading of registered or listed sites is as follows:

Grade I   Sites are of exceptional interest

Grade II*   Sites are particularly important (more than special interest)

Grade II   Sites are of special interest, warranting every effort to preserve them

Inventory of Gardens and Designed Landscapes in Scotland

Historic Scotland manages the Inventory which was first published in 1987. It is a list of nationally important sites that meet the criteria published in the Scottish Historic Environment Policy 2009. Whereas English Heritage grades the listed sites, Historic Scotland does not but awards merits: outstanding, high, some, little and none. The same planning consultation process and consent is required for any site on the Inventory.

Register of Landscapes of Historic Interest in Wales

Cadw (the historic environment service of the Welsh government), in association with the Countryside Council for Wales and ICOMOS (International Council on Monuments and Sites), compiles this register. The sites are classified with gradings which mirror those by English Heritage. The consultation process on planning applications is only voluntary, with all authorities being asked to refer applications to the Garden History Society. Grade I and Grade II* sites should be referred to Cadw.

Owners

Every owner is different and has a unique view of what should happen with his garden or park. If it is not listed on a register it can be assumed to be unimportant enough that changes, including full destruction, can happen without reference to any authority. It is more likely that the owners will be considering what they believe to be improvements and in undertaking such changes may cause an unrecognised garden to be downgraded, perfectly legally. With listed gardens the situation is slightly more complicated.

The government, under the auspices of the three heritage bodies, owns and maintains some properties that are listed on the registers. They also undertake restorations and award grants to sites they deem worthy of conserving.

Charitable organisations, usually having a not-for-profit status, own and maintain properties just as do the heritage bodies. The National Trust (founded 1894) and the National Trust for Scotland (since 1931) both own and manage a large number of parks and gardens, but sometimes a local charitable organisation, such as the Painshill Park Trust, will be established to do the same job. The National Trust Acts grant the Trust the unique statutory power to declare land inalienable. This prevents the land from being sold or mortgaged against the Trust's wishes without special parliamentary procedure. There is no prescription for how conservation, preservation or restoration must be achieved.

Discussion

English Heritage embarked on a series of restorations and reconstructions at the end of the twentieth century, starting with the Privy Garden at Hampton Court

Palace. Although as accurate as it is possible to be, many judge the result to be of lesser quality than the garden it replaced. At Kenilworth Castle they invested £2 million in reconstructing the early Elizabethan garden, as discussed in Chapter 1. Not all garden history experts are convinced that the proportions and detailing are as accurate as they could be and some have questioned the wisdom of committing such a large sum to a single project. The restoration at Wrest Park, in Bedfordshire, which was outlined in Chapter 2, represents an about-turn by English Heritage, who were originally going to focus on the eighteenth-century character by removing all the Victorian 'improvements'. As indicated earlier, the consequence is that Wrest Park will not again be how the original designers intended, nor even as English Heritage were thinking was the most worthy phase of its history, but a garden that shows elements from each period fused into one in an honest evocation of the very nature of gardens: that they are always growing and therefore changing. Whether English Heritage is right to adopt this policy is a matter for discussion. If landscape gardens are considered to be art, can they seriously remain so if they are overworked by later artists? With architecture it happens regularly enough that a great house is refaced to suit the prevailing taste and perhaps also remodelled for practical reasons. However, such buildings are rarely held in the same regard as those original ones which manage to survive unscathed. There are so very few gardens that can claim to be in their original form and those that fall into this category offer a unique window on the past.

More restorations are planned by English Heritage.

The National Trust own the overwhelming majority of sites mentioned in this book, many having come into their ownership soon after the Second World War as a consequence of overburdening death duties. Tim Richardson, writing in *The Telegraph* (14 June 2011), refers to a National Trust property, Studley Royal, where a new visitor centre was installed during the 1990s. He says:

> There were moments of sheer madness at the National Trust in the Nineties, such as the decision to site the visitor centre at Studley Royal at the wrong end of the garden, so that visitors now walk through the abbey [Fountains Abbey] – the picturesque climax of the garden – first instead of last.

W. C. T. Walker, the Assistant County Architect of North Yorkshire County Council in 1976, confirmed that the intention was always to have Fountains Abbey as the culmination of the sequence. Writing in *Landscape Design*, August 1976, he said: ' Fountains Abbey is the unforgettably majestic culmination, to be seen and appreciated only from a distance, ideally from the Surprise View above Half Moon Pond'. He continues: 'the layout of Studley Royal is specifically designed to be experienced from the beginning of Canal Gate and not from Fountains Abbey'. Before the new visitor centre was built, the car park and starting point for the garden tour was by the lake and the route progressed in the same sequence that the garden was designed in, namely upstream. As everybody except the National Trust knows, the picturesque ruin of Fountains Abbey is the climax of the sequence. It is slowly revealed as the Half Moon Pond is left behind and the bend of the river straightens out, trees either side acting as a visual framework for the ruin. To locate the visitor centre in woods away from Studley Lake is not a problem in itself, indeed it has considerable merit. The problem lies in the advice given to visitors once they arrive. For those with only two hours to spend, they suggest a stroll from the visitor centre to the abbey and the abbey mill, from where 'in summer you can enjoy an ice cream from the abbey kiosk'. If the visitor has half a day to spend, then he should first go to Fountains Hall for exhibitions, then to the abbey with a guided tour, then finally take the path to the water garden. 'Back at the visitor centre you can enjoy refreshments in the restaurant and find a souvenir of your visit in the shop.' The whole-day itinerary suggests spending more time at the mill and the porter's lodge 'to find out what it was like to be a monk in medieval England'.

Studley Royal and Fountains Abbey was awarded World Heritage Status, strictly speaking Site Inscription, in 1986. This is an onerous process typically costing hundreds of thousands of pounds and taking several years to complete. In the UNESCO Statement of Significance, reference is rightly made to the combining of the various buildings, the deer park and the water gardens into one harmonious whole. For the reasons outlined above, nobody who follows the National Trust's recommendations will experience the whole site as intended. Truly this is madness and the blame can be laid squarely at the feet of the decision makers at the National Trust. Fortunately, no part of the garden has been damaged, only the experience of it. If heads didn't roll, they should have because the Studley Royal episode did nothing for the credibility of the National Trust, the majority owner of Britain's garden and park heritage, or for the nation's trust in them.

Following the success of their work at Ham House, where full documentation allowed no room for conjecture in the restoration, the National Trust started the process of drawing up management plans for all the parks and gardens in their care during the 1980s. John Sales, their Chief Gardens Adviser at the time, recognised the need to deal with each site on its own merits, with a management regime that provided a broad outline rather than a fixed blueprint for the future. Writing in *Landscape Design* in February 1988 he said:

> Today it is essential that those in charge should understand the motives which lay behind the making of the park or garden, not just the general cultural ideals of the time but the specific ideas of the individual(s) who created it or who may have altered it since. Without this sensitive understanding the garden or park is liable to revert to a stereotype of the period.

Read in conjunction with what happened at Studley Royal, it appears that those in charge might have lost their way. In the same article he addresses the issue of management relating to periods of historical significance:

> Emphasis on a certain period in the history of the property is right where this is overwhelmingly the most important time in the existence of that park or garden but replanting as nearly as possible according to the original merely turns the clock back: the clock does not stop and the process of development continues afresh.

Whilst these sentiments seem to make perfectly good sense, I will return to them when discussing Blenheim Palace later. There would, I imagine, be widespread support for his thesis that ultimately gardens and parks are things to enjoy, 'giving delight and satisfaction, surprise and repose, according to the ideals of their creators'. He uses this to strengthen his position on 'using our judgment', suggesting that it is important to remain creative if the garden is not to die, spiritually if not physically. Sales puts his finger on the problem here. Landscape design is a unique art form in that it brings the fourth dimension into the equation. Written only months after the hurricane of mid-October 1987 when there was massive destruction of the landscape structure in parks and gardens across the country, it is understandable that he should have been concentrating on continuity and perhaps the devastating winds were a wake-up call for all those charged with managing parks and gardens.

In the National Trust's leaflet *The National Trust and gardens* (1980, revised 1983), they state that 'it is essential that someone should keep a constant eye on the whole pattern of gardens to see that each pursues a course of development right for its own particular history and character. Without this general view, gardens are liable to drift along paths dictated by the personal preferences of those in charge and the gardening fashion of the day'. This is easy to support without qualification.

Where a body representing multiple sites, as with the National Trust, can devise laudable policies applicable to their complete portfolio, private owners have a focus on one site only. The Lutyens and Jekyll garden at Folly Farm used to have eight gardeners to keep it in trim in the early years but under the stewardship of its owners in the 1970s and 1980s this had reduced to two. Vernon Russell-Smith advised the Astors on how to 'simplify without changing the feel of the garden'. The simplifications involved replacing grass with paving in the entrance court, reducing the size of borders in the barn court, making a white garden from the original hornbeam arbour, removing flower borders on the Dutch front 'canal' side, reducing in scale the flower parterre, and removing flower borders through the orchard (replaced by an avenue of crab apples). Spring-flowering bulbs were introduced to the lawn, with the intention of bringing colour from January to May. Waterside planting has also been added on the banks of the stream which marks the western boundary of the garden. So subtle and modest are they that a visitor not knowing of these simplifications, but aware of the Lutyens and Jekyll style, might not appreciate that they have been made. The difference to the maintenance budget must have been so nominal that the changes can hardly have been worthwhile financially.

Hestercombe was in need of radical restoration in the 1970s when Somerset County Council embarked on their labours. Was it fortunate that the original plans were found and allowed an accurate restoration down to the last daisy, or is this turning the clock back and stopping the process of development, as the National Trust would see it? It seems clear to me that using Jekyll's plans was the only sensible way of tackling this restoration and if the garden is now at a stage of arrested development, so be it. There was no water in the east rill because the puddled clay had cracked. Restoring it using the same detail was not practicable but restoring the water flow by modern techniques was surely more important than leaving it as a relic of a failed past.

At Claremont, the National Trust restoration of what had become known as Claremont Woods, because of the self-seeded sycamores and birches, established the mid-eighteenth century as a restoration date, thus justifying the removal of Brown's planting. However, as Graham Stuart Thomas reports in his article for *Landscape Design* in February 1979, a camellia house and a viewpoint were added in the nineteenth century and these, along with 'all important additions of the nineteenth century have been kept'. He doesn't say why and on the face of it there seems no sense in establishing a restoration date and then being selective about what subsequent elements to retain.

In the 1980s, when the National Trust was well into a programme of conserving its gardens and parks, the Duke of Marlborough commissioned Hal Moggridge and Cobham Resource Consultants to survey and prepare a management plan for Brown's landscape at Blenheim Palace. The original eighteenth-century plans identified where individual trees and clumps were planted, which, on further study, revealed a complex alignment of prospects across the park. The plan proposed re-establishing the original planting of the 1760s and to manage this over a period of 220 years. What else could it have proposed and how could any judgment not to faithfully reproduce Brown's plantings be justified? Since then the entire site has been elevated to Site Inscription as a World Heritage Site (1999) which brings with it a requirement for a management plan. The tree planting in the High Park, west of the palace and lake, was found by the consultants to be in a completely different condition from how it was intended and how contemporary paintings showed it at maturity. Then, there were grazing cattle and deer that cleared out all the understorey vegetation and allowed views under the tree canopies as well as letting in the light and giving an attractive balance of light and shade. With the grazing animals removed, dense thickets developed. There is a place for dense thickets but they don't belong in one of Brown's parks. In the early twentieth century the then Duke commissioned Achille Duchêne to make elaborate water parterres between the palace and Brown's lake on the west side and an Italian garden on the east side. These are arguably as controversial as was Brown's dismantling of Wise's massive parterre garden on the south front. It's hard to imagine any of these changes having happened after the introduction of listing, whether the site was private or not. There are still purists who lament the loss of Wise's parterre garden, and others who can't find any peace in the Italian garden.

Is Heligan a new garden or a restored one? It may more accurately be a reconstructed one. Tim Smit, before he conceived of the Eden Project, 'discovered' the Victorian garden on the other side of St Austell, near Mevagissey in Cornwall, and set about uncovering it from its overgrown and severely neglected state. Heligan, pronounced 'hLIGn', was a working estate garden with twenty-two gardeners in 1914 and the character was typical high Victorian Gardenesque although there had been earlier gardens in different styles in previous centuries. Of course there are numerous rhododendrons and camellias and the 'Lost' gardens are now, twenty-one years after they were rediscovered, one of the country's most popular garden destinations. The intention is that Heligan should remain a living and working example of the best of past practice. The success in capturing the public's imagination lay in some early television coverage and the branding of the name 'The Lost Gardens of Heligan'. Smit is a successful promoter of his projects and at Heligan he has made something out of nothing. It is unlikely that anything of any great design consequence ever existed there but the public appeal of massed rhododendrons is always hard to eclipse. There's something about India. Its influence in the kitchen has promoted Chicken Tikka Masala to the British national dish and we can thank the temperate Himalayan foothills, where many of the plants for the rhododendron-rich Victorian and later plantsmen's gardens come from, for supplying the raw materials for the default favourite of millions of British gardeners.

Lessons

The National Trust and independent observers alike have recognised that there was, at the end of the twentieth century, a flourish of garden restorations and recreations. Tim Richardson, who describes himself as an irreverent commentator on all matters pertaining to gardens and who is a determined pigeonholer of design trends, suggests that it was a phenomenon of the 1990s and that we might now be living in a post-restoration age. However, the National Trust and others were engaged in these activities from at least the 1970s and the process does not appear to have stopped yet. Richardson applauds English Heritage for its change of plan at Wrest Park and he implies that many National Trust works might have been conceived and implemented a little too hastily, in a way that will betray their status in the years to come. John Sales presented a similar view in his *Landscape Design* article of 1988: 'If we can recognise Victorian and even 1930s reconstructions of historic gardens, it is reasonable to suppose that future generations will be able to recognise the style of restoration current in the 1980s'. That would be a shame as it suggests that in the restoration of wildly differing sites there has been a corporately applied gloss, or maybe a dull eggshell might be a better simile. What would be more appropriate is for no application to be made, or if absolutely necessary, at least it should be invisible.

Jenny Little, a practising landscape architect and university tutor in 1998, wondered (*Garden Design Journal*, spring 1998) whether the twentieth century had produced any gardens worthy of preservation. Assuming this to encompass conservation, it's a good question, at least post 1930s. How, in another hundred years time, the last half of the twentieth century's gardens and parks will be viewed is anyone's guess. Compared with previous centuries there is a notable paucity of choice but that is largely due to the professionals finding most of their work in the public and corporate sectors rather than in private gardens. The Society of Garden Designers was only established in 1981 and slowly moved to embrace professional status, its journal setting an excellent standard. The country's historic garden heritage was widely covered in the Landscape Institute's journal during the 1960s, 1970s and 1980s, but interest has tailed off in the last twenty or thirty years as more focus is now placed on contemporary and global design, where most of the work now is. There is almost a requirement for designs to be avant-garde in order to command journal space. By implication, the professionals that are noticed are those producing distinctly self-conscious designs which look jazzy or different in some way – very different from the light touch of the immediate post-war years when it was sufficient to work *with* the landscape by reducing the impact of development. Twenty-first-century gardens and landscapes of the unexpected will take their place in history and only with time will the guardians of our heritage be able to make reasoned assessments of their worth.

As things are today, despite listings, gradings and planning consents, the condition of any garden or park may be altered from one visit to the next, so its value as an item of heritage may be either reinforced or downgraded. It would be desirable for management plans to recognise not only that landscapes change but also that the essence of why a site is being conserved will be its value as a historic reference point. The starting position for management should therefore be not to touch, or at least not to change, anything unless there is an imperative to do so. The dilution of designs has already happened in the years before controls and statutory planning, so much so that most of our heritage parks and gardens have been overworked and the scholar is left to pick away at the present state to discover the site's true origins. Artwork should not be treated in this way. Only by understanding the past can students appreciate the present and plan for the future, the best way of doing which is to see for themselves and not have to rely wholly upon textbooks for second-hand information. As I hope this volume has shown, the parks and gardens mentioned here are not only historic reference points but they display the essential techniques of the design process which can be applied to any project. First, take your Grand Tour.

Not all gardens and parks mentioned in this book are open to the public. Directories and local guide books should be consulted to establish visiting opportunities and private property should be respected at all times.

FURTHER READING

Apart from having visited and studied at first hand the majority of the gardens and parks described here, I have found the following volumes of particular value either as a student myself or later in my role as an educator. Of course there are hundreds of others, some general and some more focused, which were not available to me or which cover similar ground and have not been included because of this. I only offer these as a representative collection, which together can broaden the range of historical facts and figures that are included in this volume, sometimes also enlightening the reader on the techniques used by the designers.

GENERAL

A History of British Gardening, Miles Hadfield, 1969. Spring Books (Hamlyn)

This volume is a comprehensive, chronologically arranged history which leaves almost no stone unturned. Hadfield explores, in more depth than can sometimes seem relevant, everything from Roman Britain to 1939 and if there is room on your shelves for just one study of this subject, then this should be it. For its size and scope, there is perhaps too little illustrative material. Curiously, he glosses over the entire public parks movement in little more than a single paragraph; neither does he explore the Edwardian Arts and Crafts gardens in any depth.

Garden Design, Sylvia Crowe, 1981. Gibson Packard

This revised version includes more illustrations than the original, which was published by Country Life in 1958. The first section of the book includes a thoughtful survey of garden history, although her analysis of the twentieth century is a little strange. On Lutyens and Jekyll, she says: 'The partnership resulted in some very pleasant gardens of distinctive form', and continues, 'its weakness lay in its lack of unity'. Further, she concludes that the Sissinghurst, Hidcote, Tintinhull type of garden succeeds because the architectural form of the enclosures is strong enough to contain the plants. Sylvia Crowe was a thoughtful and leading landscape architect of her time but how she can characterise Lutyens/Jekyll gardens as having a lack of unity and, by implication of her comments on Hidcote, having weak architectural enclosures is simply beyond understanding.

Scottish Garden Buildings, Tim Buxbaum, 1989. Mainstream Publishing

Despite its title, this book is essentially about gardens which happen to have buildings in them, which is most of them. Buxbaum writes clearly and succinctly and illustrates his text with photos, plans and sketches.

The English Garden in our Time, from Gertrude Jekyll to Geoffrey Jellicoe, Jane Brown, 1986. Antique Collectors' Club

Brown can just about be forgiven for finding technical achievement in an 'elegant gravel sculpture' at the International Garden Festival, Liverpool 1984, where there is neither elegance nor sense, because she correctly assesses the event as being composed not of gardens in the traditional sense of a peaceful refuge, but stuffed full of theme gardens. She is also one of the few writers to recognise Michael Branch's contribution to garden design in the late twentieth century.

The Garden in England, Vernon Gibberd, 1981. Dinosaur Publications

This is a very modest little book that I found in a National Trust shop in the 1980s. Gibberd starts with Roman Britain, so is more ambitious than some, and illustrates each of the major changes with his own well-chosen drawings, finishing with a bleak scene in which a lone standard tree fights for its life in a desolate greensward populated by tower blocks and street furniture.

The Scottish Countryside – Its Changing Face 1700–2000, Rosemary Gibson, 2007. John Donald

A wonderful collection of old hand-coloured plans, maps and drawings are the nucleus of this study, which has a wider remit than designed gardens and estates but, as so many of them were so extensive, their impact on the landscape was often very significant. The plans prove this, graphically.

The Shell Gardens Book, Peter Hunt (editor), 1964. Phoenix Rainbird

Organised into sections which describe different styles and features, with a useful preamble from Miles Hadfield providing a historical overview and a good chapter on *People and Gardens* which offers brief biographies of the chief protagonists. There is also a section listing significant gardens in the UK and Ireland, by county, which is not necessarily up to date given the publication date of my own volume. Illustrations are not very helpful but the text is uniformly worthy.

Yesterday's Gardens (National Monuments Record photographic archives), Alastair Forsyth, 1983. Royal Commission on Historical Monuments, HMSO

A short introduction is followed by photographs in black and white taken at the end of the nineteenth and beginning of the twentieth century. They demonstrate how predictably conventional and dull were the mainstream gardens of the period, not least the passion for rock gardens and elaborate Victorian-style display. As such, with one or two exceptions, they emphasise the quality and invention of the gardens not included and perhaps also indicate how, in each period of history, only relatively few are worthy of merit.

33rd IFLA World Congress – Proceedings, 1996. International Federation of Landscape Architects

The theme for the Congress in Florence was *Paradise on Earth, The Gardens of the XXI Century* and the Proceedings were published in two volumes.

EARLY RENAISSANCE EXPERIMENTS

The Renaissance Garden in Britain, John Anthony, 1991. Shire Publications Ltd

Covering the early Tudor to late Stuart period, i.e. early 1500s to late 1600s, Anthony writes with considerable authority, helpfully organising the text into succinct paragraphs each dealing with identifiable subsections and individual gardens, whilst avoiding unnecessary and scholarly detail. An excellent little volume that deserves a place on all garden history lovers' shelves.

THE GRAND MANNER

Atget's Gardens, William Howard Adams, 1979. Doubleday

An important photographic record of the French gardens of the Grand Manner, seen through Eugene Atget's lens at a time when their original purpose was a distant memory and their present one, as honeypots for tourists, was yet to be imagined. It shows the gardens in partial decay with nature just beginning the process of reclaiming what had been designed to dominate her. Apart from their value as a record, they are now recognised as artwork in their own right and together splendidly convey the spirit of the style, albeit without the essential ingredient of the massed and colourful courtiers who would once have populated the gardens. The introduction is by Jacqueline Onassis, who writes with informative, almost poetic, insight.

Men and Gardens, Nan Fairbrother, 1956. Hogarth Press

This book is also available in a reprint version published by Lyons & Burford in 1997. Fairbrother writes in a charming style and appears to get distracted along numerous alleyways in the progress of her thesis. However, the strands of her slightly rambling discussions ultimately come together to reinforce both her arguments and observations. There is a particularly good chapter on what she calls *Les Jardins de l'Intelligence*, being an appreciation of the French Renaissance style.

The French Garden 1500–1800, William Howard Adams, 1979. Scolar Press Ltd

This book deals exclusively with France so might seem to be out of place in a book about British garden design. It is included because I've found no other source which provides such a good background to the rise of the Grand Manner as well as a description of the style itself, which of course was imported into Britain. The detailed analysis of the ultimate example of the style, Versailles, with plans showing the development of the design, is essential to the understanding of the Renaissance approach.

The Theory and Practice of Gardening, John James, 1712 (being a translation of A.-J. Dézallier d'Argenville's original of 1709). ECCO Print Editions (reproduction from British Library)

The book is divided into roughly equal sections dealing with the theory and the practice of gardening. Supplementing the text, which describes the Grand Manner approach, are illustrations of parterres and steps as well as techniques for levelling slopes into terraces and for setting out on the ground certain geometric shapes. This was the oracle for designers working in the time of Bridgeman.

The Romantic Revolution

Capability Brown in Northumberland, Peter Willis, 1983. Northumberland and Newcastle Society, reprinted from *Garden History*, the journal of the Garden History Society

Contains interesting plans and drawings all attributed to Brown and referring to Kirkharle, Rothley Lake and Alnwick Castle. The plans are particularly interesting in that they show his mature style. Dorothy Stroud's *Capability Brown*, originally published in 1950 by Country Life, remains the authoritative work. Chatto & Windus have recently published Jane Brown's exhaustive *The Omnipotent Magician* (2011), which *Gardens Illustrated* has called 'the first major study of Brown for more than a decade'. Both offer a great deal more detail of his life and times in addition to a detailed exploration of his big idea.

The English Landscape Garden, Miles Hadfield, 1977. Shire Publications Ltd

Hadfield is the oracle when it comes to the history of designed landscapes in

Britain. In this slim volume he summarises the movement, dealing with its origins, some of the seminal works and the different approaches of Brown, and Repton, as well as Richard Payne Knight and Uvedale Price, both responsible for promoting the Picturesque. He also declares his hand as a Brown detractor, holding him responsible for destroying the great formal gardens of London and Wise.

The Landscape Garden in Scotland, 1735–1835, A. A. Tait, 1980. Edinburgh University Press

This is a comprehensive and scholarly study of the period described and has many useful illustrations. The appendices and index are not entirely accurate, which can be frustrating, but for anyone interested in the peculiarly Scottish perspective of landscape gardening, with its great potential for the Picturesque, this volume is mandatory reading.

SCIENCE, DISPLAY AND PATTERN BOOKS

The Victorian Country House, Mark Girouard, 1979. Book Club Associates

There are good chapters on Osborne, Cragside and Standen as well as insights into architects including Shaw, Devey, Webb and Ernest George. Unlike Aslet's companion volume, there is no review of the gardens of the period but the text makes frequent references to them.

The Wild Garden, William Robinson, 1870. Scolar Press reprint of 4th ed., 1979

Robinson loved plants and every page of this book exudes his enthusiasm for them. *The Wild Garden* set him against the prevailing fashions for carpet bedding and the geometric excesses of the Italianate style as well as Blomfield's classical revival. As well as a mass of horticultural good sense, the book contains some very fine wood-engraved illustrations.

ARTS & CRAFTS

Colour Schemes for the Flower Garden, Gertrude Jekyll, 1982. Antique Collectors' Club

Originally published by Country Life in 1908, this is probably her most ground-breaking book and is organised around a calendar approach, her idea apparently being that colour need not only be the preserve of summer. Curiously, she publishes detailed planting plans from her own garden as if they could be copied to any other situation, contrary to the Arts and Crafts ethos of design being site specific.

Gardens of a Golden Afternoon, Jane Brown, 1982. Allen Lane

At the beginning of the 1980s there was a resurgence of interest in the houses and gardens of the Surrey School. The Arts Council of Great Britain staged an enormously popular exhibition in 1981 on the work of Lutyens and the Architectural Association had a complementary exhibition the following year, 'Miss Gertrude Jekyll, 1843–1932, Gardener'. This book by Brown followed very shortly afterwards, concentrating on the gardens that Lutyens and Jekyll made together. Jekyll's own books were reprinted by the Antique Collectors' Club and several authors published works on Lutyens and his country house practice, all in the early 1980s. Brown's is the only one that recognises that the gardens are just as important as the houses, sometimes much more so. Hers is a reasoned and sensitive analysis of the gardens but it should be remembered that it is her view only. In my correspondence with her we failed to find common ground on the merits of several gardens including Lindisfarne Castle, Abbotswood and Plumpton Place.

Houses and Gardens, M. H. Baillie Scott, facsimile edition of 1906 publication by George Newnes Ltd

For those who are seriously interested in Arts and Crafts architecture as well as many well-composed and finely detailed gardens of the period, this compendium of his earlier works by the author will be a fascinating read. Baillie Scott doesn't hold back in the territory he knows best but is more balanced in his discussion of gardens and sits on the fence 'in the quarrel between naturalists and formalists'. Lavishly illustrated with photographs, sketches and detailed plans, his design rationale is always easily followed.

Lutyens, Architectural Monographs No. 6, Peter Inskip, 1979. Academy Editions/St Martin's Press

Although this concentrates on the architecture, it is difficult to draw the line sometimes between where that ends and the garden begins. There are excellent colour illustrations, house and garden plans and an outline of the architect's career. The main body is devoted to analytical case studies.

New Place, Haslemere and its Gardens, W. B. Duggan, facsimile edition of 1921 publication that was printed for private circulation

It's not often that a house and garden of such relatively modest scale is written about in a volume that is solely dedicated, as here. Voysey's garden is described in some detail but the house is dealt with only very superficially.

The Gardens of Gertrude Jekyll in Northern England, Michael and Rosanna Tooley, 1982. Self-published

Tooley identifies nineteen gardens in northern England where Jekyll either advised or was commissioned to make designs, only one of which she is thought to have visited. 'A designer of advancing years and very poor sight can scarcely be accused of an abdication of responsibility for being unable to visit the gardens for which she had received a commission for a new design', he concludes. Well, yes, a designer perhaps ought not to take up a commission under these circumstances. It is little wonder that not all Jekyll's planting schemes, still less her garden designs, were successful. The Tooleys are self-confessed Jekyll hunters and write with obvious admiration for her planting skills.

The Last Country Houses, Clive Aslet, 1982. Book Club Associates

Aslet includes a splendid review of Edwardian gardens as a postscript to this book, detailing the influences that led to change. In particular he quotes Blomfield from his *The Formal Garden in England*, 1892, and the passage is worth repeating here: 'The question at issue is a very simple one. Is the garden to be considered in relation to the house, and as an integral part of a design which depends for its success on the combined effect of house and garden; or is the house to be ignored in dealing with the garden?' He goes on: '[I]t is evident that to plan out a garden the knowledge necessary is that of design, not of the best method of growing a giant gooseberry'. This was written when Blomfield was still strongly under the influence of the Arts and Crafts Movement and predates most of his classical terraces.

Modernism and Abstraction

Gardens in the Modern Landscape, Christopher Tunnard, 1938. The Architectural Press

This book contains the author's self-devised approach to designing with the architecture of the Modern Movement and International Style, and represented a significant break from the cosiness of Edwardian designs, which were still admired by the public and in some cases ridiculously appended to Modernist buildings, resulting in the stylistic clash of all time. Tunnard is perhaps not always a convincing writer but the content of this book was written when he was still in his twenties and devoting his main efforts to building a practice. After nailing his reputation to the mast he quickly went away to a safe place and to relative obscurity. His suggestions for the redevelopment of Claremont Park and the details of his ideas for 'minimum' gardens and communal gardens all have the uncompromising over-planned and uniformity-based influence of Le Corbusier. This volume was never on any of my book lists for students, but perhaps I should have been more broad-minded.

Plantsmen's Gardens

Room Outside, John Brookes, 1969. Thames and Hudson

Arguably this publication had more influence on those seeking inspiration for their small gardens than any number of visits to Hidcote or Sissinghurst. Brookes offers a digest of history in the first fifteen pages, starting with 1400 BC and finishing at about 1950. Excellent garden plans and diagrams accompany the text and, although many of his suggestions for designing small gardens now seem dated (pierced concrete screen walling, softwood pergolas painted white), the book is full of practical ideas and technical solutions. Brookes' work was heavily influenced by modern art, specifically Mondrian.

Sissinghurst, The Making of a Garden, Anne Scott-James, 1975. Michael Joseph

This does exactly what one would expect of it, in great detail, explaining the design philosophy that Vita and Harold agreed on before they started the garden. Contains some fascinating old photographs.

Vita's Other World, Jane Brown, 1985. Viking (Penguin Books Ltd)

This is subtitled 'A gardening biography of V. Sackville-West' and has considerably more than the student of garden design history would need for an understanding of Vita and Harold's contribution to the world of garden making. It is a comprehensive study of relationships with family, friends and lovers interwoven with garden stories of Knowle, Long Barn (where Lutyens helped with the design) and, of course, Sissinghurst.

INDEX

Windfarm Visualisation

Perspective or Perception?

Alan Macdonald, RIBA

The author draws together a blend of knowledge and experience to explain the many scientific disciplines involved. He gives an overview of how some simple fixed standards facilitate proper validation and testing to restore confidence in visualisations which allow realistic prediction and effective planning.

- *This book should be on the desk of every planning officer and should also be made readily available to community councils, user groups and others with a genuine interest in the ever-growing impact that windfarms are having on our upland landscapes.* **Roger Smith, The Great Outdoors**

- *An excellent text for those involved in the windfarm planning process and to educate the public, I found this book also has significant applicability to 3D professionals, urban planners, architects and engineers involved in visualisation. ... this book contains a wealth of information about evaluating windfarm visualisations. It explains the methods being used ... and how anyone can evaluate them properly and knowledgeably. The explanations and discussion in this book are richly supported with excellent graphics that support the text well.* **Jeff Thurston, 3D Visualization World**

ISBN 978-184995- 053-4   280 × 210mm   144pp   liberally illustrated   hardback   £75

Patrick Neill

Doyen of Scottish Horticulture

Forbes W. Robertson

This engaging book contains a wealth of historically valuable observations and also an insight into Edinburgh's scientific scene in the early 19th century. Patrick Neill is revealed as one the most interesting Scotsmen of the 19th century in terms of the variety of enterprises he fostered and the friendships he enjoyed with so many natural scientists of his day.

- *…excellent book describes Neill's unique place in Scottish horticulture… There are detailed accounts of Neill's friends and acquaintances and precise itineraries of his travels in Scotland, France and the Low Countries.* **Historic Gardens**
- *…I was particularly fascinated by the section on Neill's 1813 report on Scottish gardens and orchards… It contains all kinds of historical gems.* **Scotsman Magazine**
- *…A fascinating biography of a little-known pioneer.* **The Scots Magazine**

ISBN 978-184995-032-9   240 × 170mm   144pp + 4pp colour   softback   £16.99

To order any of our books, please contact:

Whittles Publishing, Dunbeath, Caithness, KW6 6EG
tel: 01593 731-333   fax: 01593 731-400   e: info@whittlespublishing.com

www.whittlespublishing.com